T0134200

WRITING IN BIOLOGY

BRIEF GUIDES
TO WRITING IN THE DISCIPLINES

Edited by
THOMAS DEANS, *University of Connecticut*
MYA POE, *Northeastern University*

Although writing-intensive courses across the disciplines are now common at many colleges and universities, few books meet the precise needs of those offerings. These books do. Compact, candid, and practical, the *Brief Guides to Writing in the Disciplines* deliver experience-tested lessons and essential writing resources for those navigating fields ranging from Biology and Engineering to Music and Political Science.

Authored by experts in the field who also have knack for teaching, these books introduce students to discipline-specific writing habits that seem natural to insiders but still register as opaque to those new to a major or to specialized research. Each volume offers key writing strategies backed by crisp explanations and examples; each anticipates the missteps that even bright newcomers to a specialized discourse typically make; and each addresses the irksome details that faculty get tired of marking up in student papers.

For faculty accustomed to teaching their own subject matter but not writing, these books provide a handy vocabulary for communicating what good academic writing is and how to achieve it. Most of us learn to write through trial and error, often over many years, but struggle to impart those habits of thinking and writing to our students. The *Brief Guides to Writing in the Disciplines* make both the central lessons and the field-specific subtleties of writing explicit and accessible.

These versatile books will be immediately useful for writing-intensive courses but should also prove an ongoing resource for students as they move through more advanced courses, on to capstone research experiences, and even into their graduate studies and careers.

OTHER AVAILABLE TITLES IN THIS SERIES INCLUDE:

Writing in Engineering: *A Brief Guide*
Robert Irish
(ISBN: 9780199343553)

Writing in Political Science: *A Brief Guide*
Mika LaVaque-Manty, Danielle LaVaque-Manty
(ISBN: 9780190203931)

WRITING IN BIOLOGY

A BRIEF GUIDE

Leslie Ann Roldan
Mary-Lou Pardue
MASSACHUSETTS INSTITUTE OF TECHNOLOGY

New York   Oxford
Oxford University Press

Oxford University Press is a department of the University of Oxford.
It furthers the University's objective of excellence in research,
scholarship, and education by publishing worldwide.

Oxford   New York
Auckland   Cape Town   Dar es Salaam   Hong Kong   Karachi
Kuala Lumpur   Madrid   Melbourne   Mexico City   Nairobi
New Delhi   Shanghai   Taipei   Toronto

With offices in
Argentina   Austria   Brazil   Chile   Czech Republic   France   Greece
Guatemala   Hungary   Italy   Japan   Poland   Portugal   Singapore
South Korea   Switzerland   Thailand   Turkey   Ukraine   Vietnam

Copyright © 2016 by Oxford University Press

For titles covered by Section 112 of the US Higher Education
Opportunity Act, please visit www.oup.com/us/he for the
latest information about pricing and alternate formats.

Published by Oxford University Press
198 Madison Avenue, New York, New York 10016
http://www.oup.com

Oxford is a registered trademark of Oxford University Press.

All rights reserved. No part of this publication may be reproduced,
stored in a retrieval system, or transmitted, in any form or by any means,
electronic, mechanical, photocopying, recording, or otherwise,
without the prior permission of Oxford University Press.

The CIP Data is On-File at the Library of Congress.

ISBN 978-0-19-934271-6

Printing number: 9 8 7 6 5 4 3 2 1

Printed in the United States of America
on acid-free paper

BRIEF TABLE OF CONTENTS

TABLE OF CONTENTS

CHAPTER 3 **Strategies for the Laboratory Report**   **96**

CHAPTER **4** **Strategies for Literature Reviews 116**

PREFACE

Biology students often picture themselves in the laboratory or in the field. What they miss is that an equally important part of their job is to convey the results of their experimental work to multiple audiences. Students must also evaluate and synthesize the published literature as a basis for contributing their own findings. In short, students need to learn to communicate and read like biologists. The goal of this book is to give them a pragmatic hand up in learning those key habits of communication.

Although there are many writing guides in biology, this one meets the need for a text that is compact, economical, and attuned to the realities of contemporary biological research. Our book focuses only on the most important genres, offering experience-tested strategies for effective communication and steering students away from the most common mistakes. At the same time, it can be used in many subfields and at many levels: students introduced to our book in a lower-level course can apply the material in later courses, graduate studies, and even their professional lives. All the while we remain true to the realities of contemporary biology research, drawing upon the authors' many years of teaching undergraduate communication-intensive courses while managing their own busy research agendas.

We introduce the book with a chapter that inventories the many genres in which biologists write and explains *how* and *why* those structures get modified for various purposes. This insider's view of the construction of biology writing helps orient students who are just beginning to read the primary literature and gives them a framework to evaluate that literature.

We go on to give students detailed, process-oriented descriptions of various kinds of writing that they may encounter during their academic training. Each chapter starts with an overview and then delivers a step-by-step tour through the process of completing of a project. Key points are illustrated with annotated examples, most from student papers, and reviewed at the end of each chapter with a checklist.

The final chapters in this brief guide present the basic conventions for style, source use, and citations. These stand-alone resources will not only serve as a handy reference but also convey the message that details and clarity are important.

In sum, *Writing in Biology: A Brief Guide* presents students a sophisticated yet accessible approach to mastering the writing strategies valued by professional biologists.

ABOUT THE AUTHORS

LESLIE ANN ROLDAN is a Lecturer in the Writing, Rhetoric, and Professional Communication (WRAP) program at the Massachusetts Institute of Technology. She teaches scientific written and oral communication, primarily in the department of Biology, and regularly presents research on how undergraduate students develop scientific communication skills. Her experience teaching and researching scientific communication has helped her identify where students need the most help: not only in identifying the content for their assignments, but organizing it as well. She is particularly keen on helping students become more aware of the decisions they make as they write. Dr. Roldan holds a BA in English from Stanford University and a PhD in Biology from MIT, and joined the MIT WRAP program in 2005. She was also a scientific editor with Benjamin Lewin's Virtual Text (2001–2005), where she commissioned and edited college-level biology articles for the web, and the Executive Director of the Cell Decision Process Center (2009–2012), an NIH Center of Excellence in Systems Biology.

MARY-LOU PARDUE, an internationally known geneticist and cell biologist, is the Boris Magasanik Professor in Molecular Biology at the Massachusetts Institute of Technology. She has published extensively on chromosome structure and on mechanisms of genetic expression in higher organisms. Professor Pardue received a PhD in biology from Yale University in

1970, and joined the MIT Department of Biology in 1972. She has served as president of both the American Society of Cell Biology and the Genetics Society of America. She is a member of the National Academy of Sciences, a Fellow of the American Association for the Advancement of Science, and a Fellow of the American Academy of Arts and Sciences. Professor Pardue has been on editorial boards of a number of professional journals. She has also served on scientific review committees for both governmental agencies and non-governmental organizations, such as HHMI and the Burroughs Wellcome Fund, and serves as a regional judge for the Siemens competition in Math, Science, & Technology.

ACKNOWLEDGMENTS

The authors wish to thank Mya Poe and Tom Deans for giving us the opportunity to participate in the *Writing in the Disciplines* series and also for invaluable advice during the development of our book. Members of the editorial staff at Oxford University Press, especially Frederick Spears and Garon Scott, have provided wise guidance throughout this project.

We are grateful to our reviewers: Deborah Begel, Northern New Mexico College; Svetlana Bochman, The City College of New York; Lori Frear, Wake Tech Community College; Christine Griffin, Ohio University; Denise Kind, University of Alaska Fairbanks; Sharon Klein, California State University, Northridge; Robert Patterson, Washington University in St. Louis; Vicki Russell, Duke University; Roxann Schroeder, Humboldt State University; Cara Shillington, Eastern Michigan University; Sheela Vemu, Northern Illinois University; Franklin Winslow, Borough of Manhattan Community College. Their perceptive advice has strengthened the final version of our manuscript.

We are indebted to our students, who not only contributed the examples used in this book, but have inspired us to improve our teaching over the years. We also thank our colleagues in the Biology Department and the Writing, Rhetoric, and Professional Communication program at MIT for providing us with a scholarly environment committed to excellence in teaching and science.

We are also grateful to Marcy Thomas and her students at Wellesley College for their contributions to the Laboratory Report chapter, as well as to Jan van Aarsen for the line drawings in the Oral Presentations chapter.

Finally, we thank our families for their support and patience.

WHY AND HOW
BIOLOGISTS WRITE

Writing is Integral to the Professional Lives of Biologists

The people we know who decided to study biology in college did so for many different reasons. Some were interested in medicine, some in animals, some in plants, and some in the environment. Several of us had read Darwin's accounts of the voyage of the *Beagle* and wished we had been on the trip. When we were students we envisioned a future doing research in laboratories or in the field, practicing medicine, or teaching. And yet for all our awareness of what biologists did, we had very little sense of how much they wrote. Our science courses put very little emphasis on writing, and few of us realized how important writing would become after we settled into our different occupations.

Looking back, the fascination of Darwin's stories should have shown us the importance of good writing about biology. Of course, Darwin belongs to a small group whose contributions have shaped our understanding of the world; but even in this elite company he stands out as an inspiration to many scientists. (Stephen Jay Gould, a distinguished paleontologist also known to the general public for his widely read essays in the magazine *Natural History*, frequently listed Darwin, his father, and the

U.S. baseball player Joe DiMaggio as the major influences in his life.) Darwin's writing made an impact on many people because it allowed them to share his intellectual journeys.

Although students may not aspire to write like Darwin, we can assume that they will, as biology students, need to write. Over the last few decades academic departments in the United States have begun integrating writing and speaking instruction into courses across the university, including biology. This trend has meant that professional biologists are now teaching writing to students like you.

Since 2005 we (Mary-Lou and Leslie) have taught an upper-level Biology course with a strong communications component. Developing and refining our *Topics in Experimental Biology* course led us to write this book, allowing us to share what we have learned about teaching biological writing. Although our students go into a variety of careers—research in academia or industry, medicine, teaching, public health, environmental protection and wildlife management, consulting, science journalism—students learn, sooner or later, that writing invariably becomes an important part of what they do. And writing will undoubtedly become an important part of what you do in your profession, within the United States and internationally.

Biology Has Many Subfields that Share a Core Writing Style

Biology is a rapidly expanding field. New findings have opened up unexpected links between old subfields of research, and technical advances have made possible new approaches to old questions. For example, the ability to sequence and assemble DNA isolated from mixed populations of microorganisms collected from deep-sea water or from the skin of your elbow has revealed an amazing number of living organisms that were not known to exist. These discoveries have led to notable advances in studies of life in the deep seas and also studies of the complex population of microorganisms living on and in our bodies. Why were these organisms not discovered earlier? Before the new sequencing technology, scientists had to isolate each microorganism and grow it in the lab to obtain enough DNA to study. Among the many problems with this approach is the difficulty (impossibility?) of culturing most of these organisms in the lab.

Studies of deep-sea organisms and the human microbiome are only two of the areas affected by new sequencing technologies. Other new technologies, such as improved imaging techniques for mapping brain activity and technology for tracking animal behavior in the wild, have made biological problems increasingly attractive to scientists from math, physics, and chemistry, and to many engineers. At MIT, the increasing interest in biology has even caused biology to be added to the General Institute Requirements, the list of courses that are considered essential for all undergraduates, regardless of their major field. Biological problems are now studied in most departments in both Science and Engineering.

The subfields of biology are many, and changing rapidly, but expectations for writing remain relatively consistent across the discipline. In this book we will focus on those core expectations and the strategies you can employ to meet them.

Biologists Write for Multiple Purposes and Audiences

Whatever their job descriptions and fields of specialization, biologists need to communicate effectively in a few key genres. According to John C. Bean, genre "refers to recurring types of writing identifiable by distinctive features of structure, style, document design, approach to subject matter, and other markers" (Bean 2011). Novice writers often think of a genre as a static format into which you plug content. While that is to some degree true, seasoned writers know that recalling a format is never enough, that we also need to consider purpose (the reason for being written) and audience (the type of people who will be reading the document).

Table 1.1 gives a quick guide to many of the genres found in biology, along with their purposes and audiences. Figure 1.1

TABLE 1.1   A Quick Taxonomy of Genres of Biological Writing

Genre	Purpose	Audience(s)
Grant application	Persuades a funding agency that a planned study should be supported	Evaluation is usually by a panel of scientists in related areas of biology
Laboratory notebook	Provides detailed records of each experiment; the records are written while the experiment is being performed	Mostly the author of the notebook, but should also be clear enough for other members of the lab to interpret
Laboratory report	Reports original research by the author(s) with conclusions and supporting evidence	Members of author's lab group, or of a collaborating lab; faculty teaching a lab course
Oral presentation	Presents research results in a multimedia format	Generally colleagues within or outside of lab
Peer review for grant applications	Analyzes strengths and weaknesses of the grant applications for funding decisions	Other scientists on the panel that makes funding decisions
Peer review for manuscripts	Advises journal editors about manuscript suitability for publication; gives authors advice for improving manuscript	The editor of the journal and the authors of the manuscript
Progress report	Describes accomplishments on a partially completed project and discusses changes in future plans based on these accomplishments	Usually an administrator of the agency or company funding the project

Genre	Purpose	Audience(s)
Research article	Reports original research carried out by the author(s), with conclusions and supporting evidence	Professional biologists, predominantly those specializing in fields related to the subject of the article
Review article	Reviews and evaluates the literature of a field or sub-field to illuminate a controversy or unanswered question	Professional biologists, those working in the field of the review and those who want an introduction to a new field
Scientific poster	Presents research results in an environment where the author can interact closely with interested individuals	Professional biologists, usually with similar areas of expertise

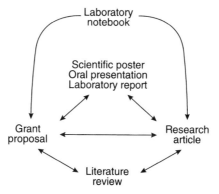

FIGURE 1.1 Cycle of Biological Writing. The genres found in biology are arranged by the way they influence the content of other genres. For example, the data found in laboratory notebooks inform the content of many genres, but published literature reviews are mainly drawn from research articles and grant proposals. Our book will describe all of these genres except the grant proposal because the formats of grant proposals vary greatly depending on the funding agency and the purpose for which funds are being requested.

shows how a subset of those genres—the most commonly used ones in our field—relate to each other. When you prepare any of these genres for a class, you may think that the purpose is to obtain a good grade from the real reader (your grader). The best way to accomplish this goal is to keep in mind the imagined purpose(s) and reader(s) described in Table 1.1.

Even when deploying a specific genre, a biologist must consider multiple types of readers. For example, the audiences for the most common document—a research article published in a professional journal—are diverse and may vary from paper to paper. Each of these audiences expects the work to conform to the values that underlie biological research: rigorously controlled experiments, reproducible results, quantification, and objective analysis. Although these values do not change, the author needs to consider different audiences when deciding on details such as the depth of the background material in the Introduction or the breadth of the Discussion. Ideally the paper will be detailed enough to satisfy a specialist in the field while still being clear enough that interested biologists in other subfields can understand the work. In other words, each genre and each paper has overlapping and changing sets of audiences. Facilitating the communication between these audiences amplifies the value of the research. Keep in mind, too, that although the major journals in biology are published in English, the audience is international. In this book, our goal is to help you read and write fluently in biology, understanding the various audiences in play and moving with more confidence and control through the genres.

The Structure of a Research Article Informs the Structure of Other Forms of Biological Communication

While we ideally want you to gain mastery of several genres, in this book we focus most on writing research articles (Chapter 2).

We opted to do this not only because a very brief guide cannot cover everything but also because research articles constitute the primary literature of biology. They are considered primary because they are written by the scientists who performed the research studies. Research articles are published in peer-reviewed professional journals. Therefore they have been reviewed for quality, accuracy, and relevance, among other attributes, by two or more independent experts (or "peers") before being accepted for publication. These reviews, together with the fact that interested readers can decide whether data in the paper fully support the conclusions, make research articles the most reliable biological literature that we have. Nevertheless, the primary literature is not beyond question: new discoveries may affect assumptions on which the conclusions were based, changed conditions may produce different results, or technical advances may provide more reliable data.

Your first meeting with a scientific research article can be intimidating. The paper contains many unfamiliar words and acronyms and is usually aimed at experienced scientists in the same subfield. Nevertheless, behind that complexity, research papers have a clear and reasonable structure. Once you understand that underlying structure, you will find it easier to navigate and comprehend any paper. In fact, experienced scientists take advantage of their genre knowledge to navigate papers outside their major field because they also need to grapple with concepts and words (and especially acronyms) that are unfamiliar.

Research articles in peer-reviewed journals reflect the logical progression of the typical research project. They begin with an Introduction, followed by Methods, Results, and Discussion. (This organization is frequently abbreviated IMRD.) In most papers these sections are separated and explicitly named. In papers where there is no marked separation, information is still usually presented in this order. The Introduction presents the question to be studied and explains why this question is important by reviewing relevant literature in the field. The Methods section provides details of the experimental approach, and the Results section presents the evidence that the authors have obtained to support their argument. The Discussion section evaluates the strengths and weaknesses of the evidence, and considers broader implications of the results and ways to extend the study.

The better you understand how the typical research article is organized, the better you will be at comprehending and evaluating published articles. Understanding the IMRD structure can also help you organize your thoughts as you begin to plan your own research projects and compose your own papers, lab reports, or literature reviews. For these reasons, we begin this book with a detailed analysis of the sections of a research article and the strategies that experienced writers deploy when composing each section. Later in the book we coach you through the other key genres that you are likely to encounter in both college biology courses and the profession. Yet even as we turn to those other genres, we often refer back to the underlying structure and core values driving the research article, which serves as a kind of parent to the larger family of genres you may need to write, including laboratory reports, literature reviews, and posters.

How to Use This Book

Understanding the structure of the research article, described in Chapter 2, will give you a running start on the other common forms of communication in biology because other genres have similar structures, and indeed help inform the content of the research article (see Figure 1.1). The laboratory report, presented in Chapter 3, might be considered a "research-article-to-be" because it discusses work that could well become part of a research article if combined with results from related studies.

However, as it stands, the work in the laboratory report is more limited in time and scope and is written for a narrower audience than a research report. Oral presentations and scientific posters present material from a research article or laboratory report in different media. These two formats are described in Chapter 5.

Literature reviews, discussed in Chapter 4, constitute what is called the secondary literature because they analyze and synthesize a set of primary research articles. Many professors ask students to do literature reviews so that they can gain some experience finding, selecting, reading, and synthesizing the primary literature—which most students, as newcomers to this kind of reading material, find quite difficult. We certainly see that kind of assignment as valuable, but as working scientists we also see the literature review in a larger framework, as somewhat analogous to an expanded Introduction section of a research article or to the Background and Significance section of a research grant application. In fact, a literature review can make a good point of departure for either a research article or a grant. Literature reviews provide entry to the literature of a subfield and help readers identify relevant research articles because the reviews discuss a number of research articles in the context of that subfield. However, it is important to remember that literature reviews, like the Discussion section of a research article, reflect the reviewer's interpretation of the data reviewed: before you cite papers that you found in a review article, you should look at the original papers to see if you agree with the reviewer's interpretation.

Regardless of the genre you have chosen (or are assigned), the quality of your work depends heavily on the clarity of your writing and the reliability of your sources. Chapter 6: *Style* will help you with effective sentence-level editing. That chapter delivers practical tips for improving the precision, readability, and credibility of your prose. The chapter also alerts you to the most common mistakes that writers make. Chapter 7: *Sources* will guide you through the process of finding, reading, and citing your sources. Every newcomer to a field finds the process of navigating the published research a bit mysterious and daunting, so we offer strategies for getting started. Even experienced researchers build their work on—and in response to—the published work of others. In either case, you must know how to mine the literature for the information that you need. You also need to know how to properly credit and cite the published work you are discussing.

We suggest that whatever the form of writing you plan to do, you begin by reading Chapter 2 about the research article to get an overview of the general model. After that, you can explore the sections on other genres, but even there you will find references to Chapter 2 that are especially relevant.

2

STRATEGIES FOR THE SCIENTIFIC RESEARCH ARTICLE

The core of biology writing is the **scientific research article**. The scientific research article has a specific structure called IMRaD (for the sections Introduction, Methods, Results, Discussion)—pronounced "im-rad"—that facilitates both writing and reading. Understanding how information is distributed among the different sections and structured within each section will help you write more clearly and persuasively for other biologists. To illustrate this point, see if you can identify the biologists in Figure 2.1, and their contributions to science.

FIGURE 2.1 Biologists Whose Writing Affected their Impact on the Field.

Chances are good that you recognize the scientists on the left: James Watson and Francis Crick. They are famous for discovering the structure of DNA. Yet without the work of the guy on the right, Watson and Crick would have had little interest in DNA.

Still no clue? The biologist on the right is Oswald Avery. Does the name ring a bell? He and his colleagues Colin MacLeod and Maclyn McCarty determined that DNA—as opposed to protein—constitutes the genetic material responsible for heritable traits (Avery et al. 1944). Why isn't the discovery of such fundamental principle better known? Some rhetoricians such as S. Michael Halloran suggest that it is because of Watson and Crick's writing (Halloran 1984). First, Watson and Crick wrote about their discovery in multiple articles (Watson and Crick 1953a, 1953b, 1953c), each oriented to a slightly different audience. Second, they understood how to use structure and concision to their advantage. Their most famous paper, "A Structure for Deoxyribose Nucleic Acid," published in 1953 in the "high-impact" journal *Nature*, is extremely short (about a page) and written in a style that makes the content accessible to other biologists, regardless of their specific expertise. In contrast, the 1944 paper "Studies on the Chemical Nature of the Substance Inducing Transformation of Pneumococcal Types," by Avery, MacLeod, and McCarthy, which was published in the *Journal of Experimental Medicine*, is wordy, overly cautious, and difficult to read. The Results section is poorly organized and contains information on Methods. Indeed, the term "DNA" doesn't appear until halfway through the paper.

So what is one lesson to learn from Watson, Crick, and Avery? Scientific writing is not merely about communicating results. Scientific writing is about conveying your results to various audiences. Learn how your science translates to the page and use scientific writing to make your writing persuasive to fellow biologists.

How to Use This Chapter

This chapter will explain how a biologist would go about developing a research article. The chapter will be somewhat longer than the others in this guide. Why? The research article is the genre that you will encounter most often when you read and conduct research as a professional biologist, whether you become a researcher in academia or industry. You will likely read a lot of research papers in your life if you become a professional biologist, so it's useful to know how they're written. Also, as we explained in Chapter 1, in learning to write a scientific research article, you will be learning skills that can be applied to writing many other kinds of biology texts.

Table 2.1 lists the purpose and approximate length of each section of a research article; the lengths are estimated on student reports, but published lengths depend upon the journal and what the author has to say. In terms of scope of content, the research article is shaped like an hourglass (Figure 2.2): the Introduction starts out broadly and narrows down to the aim of the research article; the Methods and Results sections are narrow in scope because they focus only on the project; the Discussion broadens in scope as it situates the work into the context of the field.

FIGURE 2.2 The Scientific Research Article as an Hourglass.

The sections in Table 2.1 are listed as they typically appear in a journal, but this is not the order in which professional biologists usually write a paper. Therefore we have organized this chapter in the way most papers are written. Most biologists begin by deciding on their pictures, tables, and graphs because these are directly derived from the data in the laboratory notes. Often the scientist has already prepared some of these figures and tables for use in oral presentations and posters before she has completed enough experiments to decide what the story of the complete paper will be. When the complete set of illustrations has been decided, the set can be arranged to present the experiments in the most logical order and to guide the writing of the Results section. After writing the Results, a biologist

TABLE 2.1   **Each Section of a Research Article Answers a Different Question**

Section	Question	Approx. Length
Title	What is the take-home message of the writer's research?	≤85 characters, including spaces
Abstract	What are the goal, key findings, and impact of the project?	125–200 words
Introduction	What did the writers know before beginning the project?	500–750 words
Methods	What did the writers do?	1,000–1,250 words
Results	What did the writers observe?	750 words, with five to seven illustrations
Discussion	What does it all mean?	750 words
Acknowledgments	Who helped make this research possible?	150 words

Word and illustration estimates are based on student papers that describe individual (two-semester) research projects. The actual relative lengths of the sections and number of illustrations depend on what the author wants to say and show and on the space the journal will allow.

might move on to the Methods because the section generally requires gathering nitty-gritty details; to the Introduction because she has a clearer idea of what a reader needs to know to understand the project; and then to the Discussion because elaborating on the implications is a natural way to extend the context of the Introduction and the story of the Results. Regardless of the order in which the other sections are written, the title and abstract are generally the last to be drafted because they must represent the most recent version of the research article.

Throughout this chapter, we incorporate examples of student writing. The names of the genes/proteins may be unfamiliar to you because in many cases they have been replaced by fictitious names. The point is not to describe the biology but to show how the ideas have been clearly expressed.

Figures and Tables: Start With Your Data

Figures and tables (hereafter, collectively referred to as illustrations) are the backbone of any research article because they tell the "story" of the research conducted. They help distill and organize information, and can reveal trends or relationships within the data. Ultimately, illustrations help convince your audience of your findings and of the quality of your results.

Illustrations are so important to scientific research articles that most scientists start by designing their illustrations before they begin writing their articles (of course, they have already done lots of writing in their laboratory notebooks). Many readers also look at the illustrations before reading the text—no reader wants to wait to read exciting findings. Besides, illustrations are fun. Most students, however, don't think about the importance of illustrations before they start writing, choosing instead to write the Methods first. Later, when they write the text for the illustrations, they discover that the Results don't match the Methods. Be strategic. Figure out the "story" of your research first and convey that message through your illustrations.

Even for visuals, some text is essential. Each of the types of illustrations listed in Table 2.2 needs a legend to place it in the context of the research it is supporting. Without a legend the reader is like one of the five blind men who each described an elephant differently because each felt a different part of the animal. If a legend had informed them that five people were doing the study, the men could have combined their conclusions and produced a much better description. Thus, for each illustration you must decide:

- What are the most important data to illustrate your finding?
- What is the best type of illustration to use for this?
- What information does the reader need in the legend?

In this section we will discuss how a scientist constructs illustrations by first concentrating on the principles of constructing the visual part of different types of illustrations. We will then describe general principles on which the legends are constructed.

Illustrations Convey Data

Data come in two forms, raw and processed:

- **Raw data** are data derived from your experimental procedure—for example, electrophoretic gels, or the series of numbers that are generated from an assay, observation of a behavior, or counting mutant animals.
- **Processed data** consist of data that have been manipulated or modified. Examples include graphs, presenting the number of mutant animals as a fraction of the total progeny of a cross, or quantifying the intensity of bands on a gel or the amount of signal in a microscope image.

So, what type of data should you show in your research article? Generally speaking, processed data make it easier to demonstrate trends, and numbers can be more persuasive than simple bands on a gel. Frequently, however, you will want to include the image of your gel, study site, organism, and so forth—raw data—to show the quality of the results measured to obtain your processed data.

Do You Really Need an Image?

While research articles almost always include illustrations, not everything in an article needs one. Students tend to include all the data they generated to demonstrate the amount of work they have done. But showing all your raw data isn't very effective. Instead, your job is to select relevant data and then design illustrations that best represent the "story" of your research. For example, perhaps you measured both egg hatchability and survival to fertile adult for your mutant, but further work showed that hatchability and adult survival were affected by such different aspects of the mutation that you should discuss them in separate papers. In that case, data from hatchability would be irrelevant in the paper about adult survival, and vice versa. In the process of deciding on your story, you may delete illustrations or combine them. You may also find that some information is better conveyed with just words. Samples that show no significant difference from each other, for example, can be cited simply as "data not shown."

Choose the Right Format for Your Illustrations

Illustrations (shown in Table 2.2) may be found in any section of the research article but they are most common in the Results, where they present the data that support the conclusions of the paper. Data can be presented in many formats, such as photos of gels or cytological preparations, graphs or tables of numerical data, and diagrams of molecular structures or phylogenetic

TABLE 2.2 Types of Illustrations

Type	Functions
Tables	Present numbers Show repetitive data Display raw or processed data
Graphs	Highlight trends Make comparisons between sets of data
Images	Present raw data
Diagrams	Describe a model Explain a technique Clarify the elements of an image

trees. Illustrations are much less often found in other sections of the article, although background information, such as schematic diagrams of biological pathways, phylogenetic trees, or macromolecular structures, is sometimes found in the Introduction and conclusions are sometimes illustrated by diagrams in the Discussion. Tables can also appear in the Methods section, where they are an efficient way to list the names and characteristics of multiple mutant stocks, PCR primers, and so forth. Figures are numbered and placed in the paper in the order in which they are discussed. Tables are also numbered in order of appearance but they are numbered separately from figures; their titles also appear above the illustration.

Tables

Tables are used when you want to present numbers or need to show repetitive data. They're not flashy and they're not good for showing trends, but they are useful when you want to present specific data points.

Tables are organized to help your reader evaluate information easily. Because it is easier to compare data down a column than across a row, the first column generally contains the independent variables (variables that you control, such as the strains in Table 2.3), while subsequent columns contain dependent

TABLE 2.3 **Bacterial Strains Used in This Study**

Strain	Description	Source
K12	Wild-type lab strain	Wild-type *E. coli* K12
HM22	AT984 hipA7 zde264::Tn10 dapA6	Moyed and Bertrand (1983)
UTI89	UPEC strain, an acute cystitis isolate	A. Clatworthy (MGH, Boston)
CFT073	UPEC strain, an acute pyelone-phritis isolate	A. Clatworthy (MGH, Boston)
FK1	K12 ΔhipA	This study

TABLE 2.4 **Expression of β-Galactosidase in Plasmids Amplified by WT and H147C *Pfu* DNA Polymerases**

Polymerase Reaction Mix	# Blue Colonies	# White Colonies
WT	3	800
H147C	5	400
WT "No primers"	1	23
H147C "No primers"	0	0
No DNA	0	0
pBS-SKT	336	0

← Label columns and rows with abbreviated yet informative phrases.

Place independent variables in rows, dependent variables in columns.

Align data to facilitate reading (e.g., numbers according to decimal point). If using whole numbers, right align (don't center) them with respect to each other. In both cases, center the column of data with respect to the column heading.

Minimize white space.

Abbreviations in labels are acceptable to conserve space.

Wild-type and mutant H147C polymerases were used to amplify DNA encoding β-gal. Transgene products of these reactions, plus controls ("no primer," "no DNA" and pBS-SKT encoding active β-gal) were transformed into *E. coli* lacking β-gal. Transformed cells were plated on media without IPTG. Expression of β-gal was assessed by the color of the colonies, with blue indicating presence and white indicating absence of expression.

variables (things you measure, such as the number of blue colonies in Table 2.4). Following are two examples of tables.

Graphs
Graphs are particularly useful for highlighting trends or making comparisons between sets of data. As a result, you rarely see graphs in the Methods section, but you often find them in the Results and Discussion.

There are many different types of graphs (e.g., line graphs, pie graphs, bar graphs, dot graphs, box plots), and you should choose the type that best fits your purpose. It may even be worthwhile to try out different types to see which one is most appropriate. In general:

- Line graphs (Figure 2.3) are best for trends (particularly over time).
- Venn diagrams (Figure 2.4) are best for showing overlap between two sets of data.
- Bar graphs (Figure 2.5) and dot graphs (Figure 2.6) are best for comparing values of different sets of data.

It is important to label both axes on the graph with terms that will be meaningful to the reader. In addition, labels and data points should be large enough to withstand reduction because an illustration will usually take up only a quarter the size of a page, often much less. Finally, simplicity is key. You want readers to concentrate on the data points or trends, so avoid the distractions of extra horizontal lines, a different color for the background, or unnecessary 3D renderings of 2D data.

Images
Images, such as gels and photos of cells or organisms, are used to present raw data. They are sometimes combined with processed data in a single figure to give examples of the features that have been quantified. Images may also be accompanied by diagrams to facilitate interpretation.

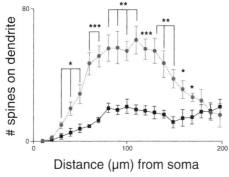

The annotations in the figure read:

- Avoid distracting elements like horizontal lines across the graph.

- Try to rely on different line styles or data symbols (here, circles and squares) rather than color.

- The placement of dependent and independent variables depends upon what best conveys the message. Here, the number of spines along the dendrite (dependent variable) is plotted as a function of the distance from the cell body (independent variable).

- Indicate the distribution of the data with error bars. In the legend, explain what the error bars represent, e.g., standard deviation, standard error.

- Use a set of asterisks to highlight the significance of the data. The asterisks should be defined in the legend, e.g., * P < 0.05.

FIGURE 2.3 Inhibition of TrkB Function Correlates to an Overall Decrease in PSD-95 Puncta (spine) Density. Neurons were treated with 1NM-PP1 (n=5, black squares) to inhibit TrkB; treatment with Bph-PP1 (n=5, gray circles) served as negative control. Individual spines budding from the dendritic processes were manually traced and counted at a given distance from the soma. ***, **, * indicate significance (P < 0.001, P < 0.01, P < 0.05, respectively) based on Student unpaired t-test; error bars=s.e.m.

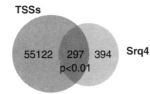

The annotations in the figure read:

- Draw circles in rough proportion to the quantities stated inside of them. Here, direct proportion would be distracting because the numbers are 2 orders of magnitude apart.

- Label circles and indicate significance of overlap, if warranted.

FIGURE 2.4 Venn Diagram Showing that Binding by the Protein Srq4 is Significantly Enriched at Transcription Start Sites (TSSs). A genome-wide map of Srq4 bound to DNA of mouse embryonic stem cells was generated by ChIP-seq. The number of regions in the genome that were enriched in binding by Srq4 (small circle) was determined by MACS algorithm, and compared to the number of such regions in a 1 kb interval around each mapped TSS (large circle). Significance of overlap was calculated by permutation tests.

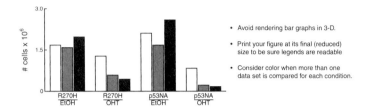

FIGURE 2.5 Restoration of p53 Expression Causes Cell Cycle
Arrest in Sarcoma Cell Lines. Proliferation capacity of
two cell lines was compared before (EtOH) and after
chemical (OHT) restoration of the inducible p53 allele,
LSL. One cell line (R270H) had one LSL allele and a
point mutation in the other p53 allele. The second cell
line (p53NA) had one LSL allele and one completely
inactive p53 allele. Cells were plated at density 10^5
cells per 10 cm dish, grown for four days, harvested,
counted and replated at the initial density. The procedure was repeated for three passages (P1, white bars;
P2, gray bars; and P3, black bars).

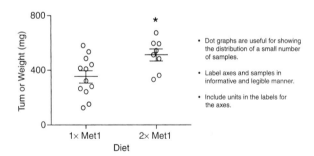

FIGURE 2.6 Metabolite 1 (Met1) Increases Tumor Growth in
Mice. Pancreatic cancer cells were injected into
immunocompromised mice. The mice were fed a
diet containing median (1xMet1; n=12) or twice the
median (2xMet1; n=8) concentration of the human
physiological level of Met1. After three weeks, the
pancreatic tumors were weighed. * indicates P-value
of 0.05.

On the web, you can find a number of free programs that can help you design your images. In our course, we request that the resolution of images be at least 300 dpi. If you plan to submit a paper for publication, it is important to check the specific Instructions for Authors of the journal you are considering. Journals differ in many details such as the file formats they will accept for electronic figures, the sizes allowed for published images, and the resolution required for color, grayscale, or line art images.

Images should be cropped and properly positioned. For a gel (Figure 2.7) wells may be shown, but avoid edges and

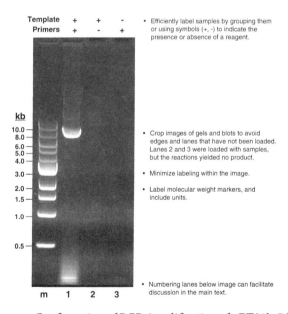

FIGURE 2.7 Confirmation of PCR Amplification of pET21b-*Pfu*. pET21b-*Pfu* was amplified by PCR using primers causing directed mutagenesis; + and − indicate presence or absence of reagents, respectively. PCR products were resolved on a 0.8% agarose gel and visualized with GelGreen fluorescent dye. The expected product size is 7.8 kb. m: 1 kb DNA ladder.

extra, unused lanes outside of the samples. All labels should ideally be outside of the image to prevent obscuring data. Label the molecular weight markers and indicate the units above the largest marker. Samples should be labeled with text rather than numbers, although numbers can be placed below lanes to facilitate highlighting specific samples in the Results section. If a consecutive set of lanes shows different combinations of the reagents present in each lane, the sample labels may be organized like a table, with symbols +/− to indicate the presence or absence of a reagent, as shown in Figure 2.7.

Microscope images (Figure 2.8) are usually cropped as well. If similar cells or organisms are shown in multiple panels or figures, position and crop each panel in the same manner. Indicate the magnification with a size bar, and, if necessary, highlight specific elements using small, discrete symbols (e.g., large arrow, small arrow, arrowhead). Written labels on the image itself are seldom readable and can obscure part of the image.

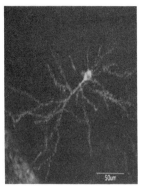

- Microscope images should include a scale bar.

- Images of similar cells or organisms should have the same magnification, size, and position when in a multipannel figure.

- If labels are necessary within the image, keep them small and discrete to avoid obscuring data. A symbol, explained in the legend, is usually better than a tiny unreadable, text label.

- Alternatively, you can add a labeled diagram next to the image.

- Explain symbols in the legend.

FIGURE 2.8   Confocal Microscope Image of a Visual Cortical Neuron Expressing PSD-95-GFP. Size bar = 50μm.

Diagrams

Diagrams are typically presented in the Discussion section to describe a model based on data presented in an article (Figure 2.9). Diagrams can also appear in the Methods or Results section to explain a technique (Figure 2.10) or to clarify the elements of a microscope image (e.g., when labels added to an image obscure the actual data). Diagrams can be more effective than photos in these cases because they give the author more control in choosing the elements necessary to convey a message. A diagram can also be used to enlarge a region of interest to indicate more detail.

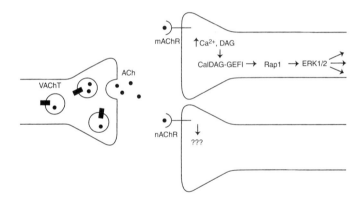

FIGURE 2.9 Model of a Proposed Neuron Parallel Pathway Mediated by Nicotinic Acetylcholine Receptors (nAChR). Mice with both *VAChT* overexpression and *CalDAG-GEFI* deletion have significantly increased stereotypy and decreased locomotion compared to mice with only *CalDAG-GEFI* deletion, suggesting the presence of a parallel pathway. We hypothesize that this parallel pathway is mediated by nAChR, but the pathway(s) downstream of nAChR are unknown.

Design Your Illustrations for Meaning

Selecting a format for your illustration is only the first step. You also want to make sure that your readers can decipher

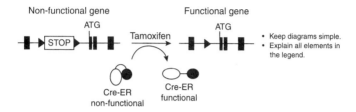

FIGURE 2.10 Construction of the Inactive Lox STOP Lox (LSL)
p53 Gene That Can be Reactivated In Vivo by
Treatment With Tamoxifen. A transcriptional/
translational STOP element (white box), flanked
by P_{lox} recombinase sites (black triangles), was
inserted in the first intron of the p53 gene, creating
a nonfunctional allele. Treatment of cells with
tamoxifen activates a Cre-ER (estrogen receptor
fused to a Cre recombinase). Recombination of the
two sites flanking the STOP excises that element,
allowing expression of p53. The p53 locus is shown
with exons as filled black boxes and introns as horizontal lines.

your illustrations. If illustrations are not easily interpretable,
they reflect badly on the rest of the paper.

One of the common mistakes that students make is not
using formatting and design techniques to help readers interpret an illustration. Students simply print out rough versions
of their illustrations and staple them to the back of their laboratory reports. Following are four design tips that will help
make your illustrations more interpretable for your readers.

1. **Consider grouping illustrations**. If you have illustrations that may be grouped together in a logical fashion,
   you may want to group some of them to create "multi-
   paneled" illustrations (e.g., Figure 1a, 1b). This grouping
   should be done on the basis of content—that is, the message, not the technique.

2. **Label illustrations in a consistent and informative manner.** Make labels that are short enough to fit in the available space yet retain enough meaning for the reader to remember what they signify. You may be tempted to refer to the samples as you did in lab (e.g., Mutation 1–5) but these numbers will mean nothing to your reader, who did not perform the experiments with you. It would be better to be more specific about your mutation (e.g., AbiE45Y or Abi delta 1-45).

> **NEEDS IMPROVEMENT**
> Mutation 1–5
>
> **BETTER**
> AbiE45Y or Abi delta 1-45

All of these labels should be explained in a legend or listed in a table. In fact, your readers will be very grateful if you use the same short and informative names throughout the text as well so that they can rapidly begin to remember the cast of characters.

3. **Minimize color.** Computer programs use color, but opting for color in your scientific articles should be an active decision, not a default. In fact, one of our former colleagues forbade his students from using color in any of their illustrations until the final paper of the course. Even then, students were allowed to add only a single color, forcing them to think very carefully about what they wanted color to do.

Why be cautious about using color? The most important reason is that too much color and detail can confuse rather than enlighten. In addition, although many scientists may first look at articles on the web, they still often print them in black and white, so any information conveyed by color may be lost. Finally, we have had students and colleagues whose red/green color blindness kept them from being able to interpret some colored figures (magenta/green is better than red/green).

We suggest a few guidelines for using color in scientific research articles:

- Use color sparingly to highlight important aspects. Color is most effective when used to highlight a specific aspect of the image. Similarly, microscope images that use two different dyes may be easier to analyze if black and white are used for both single-channel images while color is used for the image when the two channels are superimposed (to allow precise comparison of the areas stained by both dyes).
- On graphs, color can be avoided by using patterned lines (e.g., solid, dashed, dotted) or by using different shapes for their data points. Almost any shape, filled or open, will do, although avoid diamonds because they are hard to distinguish from other shapes.
- If color must be used, employ colors with very different shades (i.e., differ in the amount of black added to a color) so that they appear as clearly differentiated shades of gray when printed in black and white. If you print a color line graph on a black-and-white printer, you will notice that colors like green and blue become indistinguishable.

4. **Minimize data processing.** Digital image manipulation is now so simple that researchers can be tempted to overprocess their data when preparing figures for an article. For example, it is easy to enhance the brightness of an image to overemphasize a subtle point, or minimize something in the figure that was unexpected (e.g., a new band on a gel or an antibody-stained region in a microscope image). Although these might seem like simple

artistic changes, they falsify the data by adding to, or subtracting from, the measured signal. (Unexpected signals can lead to interesting new findings, so do not dismiss them too quickly.)

It is also tempting to crop lanes out of the image of a gel and place them in the order you wish you had thought of when you loaded the gel—or, heaven forbid, combine lanes from different gels or different exposures of the same gel. Although digital processing can easily obliterate any evidence of the new junction, journals require that the juxtaposed lanes have a solid black line in between them to indicate the rearrangement.

The best way to avoid the temptation to digitally enhance your images is to plan your experiment so that every image is of "publication quality"—clean and presentable enough for a publication. For example, load gel lanes in the order you will want in the final figure. Most of us have run into Murphy's Law (i.e., anything that can go wrong will go wrong) when trying to obtain publication-quality images right before sending out a manuscript: reagents run out, gels crack, equipment breaks, and so forth.

Data Ethics

Many journals have very strict guidelines about processing data to make sure that data are treated ethically. A good example of these guidelines can be found in the Instructions for Authors of the *Journal of Cell Biology*. Evidence that the guidelines have not been followed can result in delays in publication while the figure is redone and reviewed. In some cases, the journal may refuse to publish the article. Reusing a figure from another author without citing that author is considered plagiarism.

Write Informative Legends

Illustrations are meant to stand on their own because scientists read research articles in their subfield like a second-grader reads books: they look mainly at the pictures. In both cases the pictures are intended to carry the story. Scientists sometimes like to look at the figures and tables before reading a paper to see whether they come to the same conclusions that the author did. In most publications the images are placed close to the text that discusses the relevant work; however, space constraints sometimes require the figure to be placed on a separate page. Thus, each illustration should be accompanied by a legend that includes all the information the reader needs to interpret that figure or table so that it is not necessary to shift back and forth between an illustration and the corresponding section in the article to piece together the story. The general format of a legend is as follows:

Figure/Table number. Title. Description.

Craft the Illustration Number and Title

Illustrations are given a number that is used for reference in the manuscript (e.g., "see Figure 5"). Illustrations are numbered chronologically through the article (Figure 1, 2, 3). However, figures are numbered independently of tables (e.g., Table 1, Table 2, Figure 1, Figure 2). If you decide to move one figure, you must renumber the others.

The title is a single sentence or clause that conveys the main point of the illustration. The title may be interpretative (e.g., the conclusion or take-home point) or objective (e.g., a description of the methods or goal of the experiment).

Briefly Explain How the Data Were Obtained

After the title, the main purpose of the legend is to allow the reader to interpret the illustration. Therefore, the legend should state the experiment or the assay used for the experiment.

Interpretive Title

Figure 1. Joh2 overexpression inhibits motility of high-mobility SCC cells.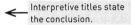

Objective Title

Figure 1. Effect of Joh2 overexpression on motility of high-mobility SCC cells.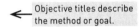

In most journals, the explanation is very brief because much of the detail can be found in the Methods. So, for example, a legend does not have to explain how PCR (polymerase chain reaction) was performed, but it does need to describe the samples tested and any other variables that were not described in Methods because such information is specific to the illustration. Samples can either be integrated in the description of the procedure or listed at the end.

For example, below is the legend you saw for Figure 2.5:

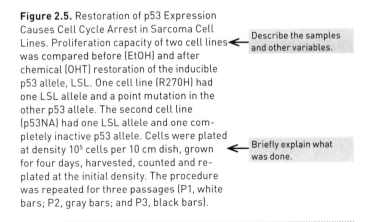

Figure 2.5. Restoration of p53 Expression Causes Cell Cycle Arrest in Sarcoma Cell Lines. Proliferation capacity of two cell lines was compared before (EtOH) and after chemical (OHT) restoration of the inducible p53 allele, LSL. One cell line (R270H) had one LSL allele and a point mutation in the other p53 allele. The second cell line (p53NA) had one LSL allele and one completely inactive p53 allele. Cells were plated at density 10^5 cells per 10 cm dish, grown for four days, harvested, counted and re-plated at the initial density. The procedure was repeated for three passages (P1, white bars; P2, gray bars; and P3, black bars).

Describe the Parts of the Illustration

To enable the reader to interpret an illustration, the legend also explains the meaning of any symbols, colors, or abbreviations added to the illustration. What do the error bars of a graph signify—standard deviations, standard errors, or confidence intervals? What do the shapes and lines represent in a diagram?

Images need a little more description because different procedures can generate similar images. Let's say for example that you show an image of an electrophoretic gel. Was the gel used to separate DNA or protein? Was the DNA detected by ethidium bromide staining or by hybridization to a radioactive probe? Alternatively, you can have a microscope image. Does the microscope image show a cell or organism? What part of the cell/organism is shown, and how is the image oriented (anterior/posterior; dorsal/ventral)?

Notice how the legend you saw for Figure 2.3 explains the various elements of the graph.

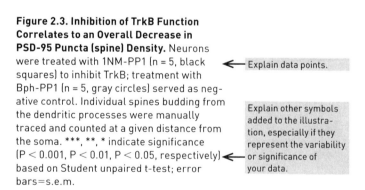

Figure 2.3. Inhibition of TrkB Function Correlates to an Overall Decrease in PSD-95 Puncta (spine) Density. Neurons were treated with 1NM-PP1 (n = 5, black squares) to inhibit TrkB; treatment with Bph-PP1 (n = 5, gray circles) served as negative control. Individual spines budding from the dendritic processes were manually traced and counted at a given distance from the soma. ***, **, * indicate significance ($P < 0.001$, $P < 0.01$, $P < 0.05$, respectively) based on Student unpaired t-test; error bars=s.e.m.

← Explain data points.

Explain other symbols added to the illustration, especially if they represent the variability or significance of your data.

Edit

In terms of language, legends are written out in complete sentences. In addition, the past tense is generally used because the legend describes how the data were obtained. Present tense, however, is used when describing symbols on the illustration (e.g., "arrow indicates the centriole", "error bars show standard error").

Sample Illustration With Legend

Although illustrations are designed to be read independently of the research article, constructing a legend requires a lot of thought. What is the background of the expected reader? What does the reader already know from reading from the paper up to this point? What details of the illustration are important for the point the author is making? Such questions demonstrate that writing a legend for an illustration demands a deep familiarity with the research article. Therefore, to fully appreciate how an illustration and legend work together, we will provide more details about the student paper from which the sample below is derived.

The sample illustration is from a study of breast cancer cells that had developed the ability to move and thus metastasize to other parts of the body. This student paper described a search for genes whose activity changed in these cells. Such genes could lead to new drug targets for cancer.

Cancers are known to be genetically heterogeneous. To separate different genotypes from the parental cell line, the student isolated single cells and grew up 30 single cell clones (SCC). These SCC lines were each tested by a motility assay: four SSC lines migrated much more rapidly than the parental population and two lines migrated much more slowly. Analysis of the RNA transcripts from these six cell lines revealed eight genes whose expression patterns differed markedly from those in the parental cells, suggesting that these genes have a role in cell motility. In this paper the student studies one of those proteins, Joh2, which is markedly downregulated in the high-mobility lines, suggesting that it inhibits mobility. To test this hypothesis, the student looked at how overexpression of Joh2 affected the migration of high-mobility SCC. Figure 2.11 shows the results of this experiment.

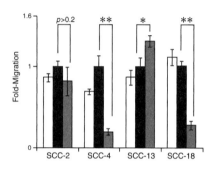

FIGURE 2.11 Effect of Joh2 Overexpression on Motility of High-Mobility SCC Cells. High-mobility SCC-2, SCC-4, SCC-13, and SCC-18 cells were transfected with a retroviral vector; the vector was stably inserted in the host genome and expressed the Joh2 gene driven by the viral LTR promoter. Motility assays for each cell line compare the infected cells (gray) with uninfected cells (white), and cells infected with a control vector (black). SCC-2 and SCC-13 cells do not show significant inhibition by Joh2: these cells may acquire increased motility by a different pathway. All fold-migration data were normalized to the cells infected with the control vector (n=3). Error bars=SEM. Single asterisks: p = 0.05 relative to control vector. Double asterisks: p = 0.001 relative to the control vector.

"High mobility"; "retroviral vector...stably inserted"; "driven by the viral LTR promoter": Information about the samples.

"Motility assays": Additional details about these assays are understood to be located in the Methods section.

"All fold-migration data were normalized"; "(n=3)": Information about the analysis for this particular experiment.

ILLUSTRATIONS CHECKLIST

Do

✓ Choose an illustration that most effectively conveys the point of the data.

✓ Write a legend that gives the information needed to interpret the illustration: describe how the data were generated, briefly list samples, and explain abbreviations and symbols.

✓ Treat your data in an ethical manner.

✓ Label objects within your illustrations in a concise yet legible and informative manner.

✓ Number tables separately from all other illustrations and in the order in which they are cited in the research article.

Don't

✓ Express the same data in more than one illustration.

✓ Include information or data points that you will not describe in the text.

✓ Use color indiscriminately: one color usually suffices.

✓ Add labels that obscure parts of the illustration.

Results: Connect Illustrations and Text

The Results section is the heart of the research article because it contains the data: the data answer the question posed in the Introduction, constitute the outcomes of the procedures described in the Methods, and provide the evidence for the argument of the Discussion section. The Results section is more than data; there is text as well. You can think of your Results section as a travelogue: you have taken pictures of your data; now you have to provide the words that make sense of the data. You need to explain why you took the photo, and highlight the aspects that make the data relevant for your argument. The Results section, then, is a descriptive narrative about your collected data.

Organize Your Illustrations to Tell a Story

For professional biologists, the Results section is scaffolded on the illustrations. Most writers start with illustrations when composing their Results, and many of a writer's conclusions are only made after seeing data plotted or gels compared.

Because the Results section is structured around the data, the first thing you need to do is to select your illustrations. A research article generally has five to seven main illustrations, although the exact range depends on the subject of the paper.

Once you have selected your illustrations, sequence them in a logical manner within the Results section. There are many ways to organize your illustrations, and the order may not resemble the order in which you performed the experiments. After all, your reader will not know (or care) about the chronological order. Think instead of how you want to structure your story. Do you want to start with your most startling result and follow it with supporting evidence? Or do you want to build your case and end with the clincher?

As an example, take a look at this set of unordered figure titles from a study that inhibited the function of the TrkB protein in mouse neurons (nerve cells) to determine the effect of TrkB on expression of another protein, PSD-95, and on overall neuron physiology:

PSD-95 neurons with inhibited TrkB function have smaller cell body sizes.

Inhibition of TrkB function correlates to an overall decrease in PSD-95 distribution.

Visual cortical neurons can be visualized by expressing PSD-95-GFP.

PSD-95 neurons with inhibited TrkB function show sparser dendritic density.

Based on these titles, it looks like the neurons were visualized by expressing the PSD-95 protein linked to GFP and the neurons were evaluated based on their cell size, amount of PSD-95, and density of dendrites. Therefore, you probably want to show first that the PSD-95 neurons could be visualized by their GFP fluorescence. But which aspect of the neurons would you show next: cell size, PSD-95 distribution, or dendritic density? Whatever order you choose (remembering the overall goal of your project will help), justify the order so that it makes sense to your reader.

Structure Your Results Section with Informative Subheads

The Results section, like the Methods, is divided into subsections, each of which is often devoted to a single illustration. Therefore, the sequence of your illustrations determines the sequence of your subsections. At this point, you can create informative subheadings for your Results section. These subheadings help readers quickly determine whether the paper is relevant to their interest. For example, based on the revised order of the figure titles above, we can construct the following subheadings:

> Visualization of PSD-95 neurons via GFP
> Effect of TrkB on cell size
> Effect of TrkB on dendritic density
> Effect of TrkB on PSD-95 distribution

The content of these subheadings does not look too different from that of the figure titles, but that's to be expected: each subsection in a Results section frequently describes only one illustration. Note, too, that these subheadings are phrases. Subheadings can also be structured as complete sentences.

Whatever format you use, try be consistent throughout your Results section.

Write Text to Narrate Your Illustrations

Now that you've got your story and a structure for your Results, writing about your data is a fairly straightforward process. The beginning of the first subsection should give a brief overview of the goal and experimental design to help the Results section be independent of the rest of the paper. Then, provide the following type of information for each illustration or major set of experiments: **rationale**, **data description**, and **conclusion**:

- The **rationale** explains why you performed the set of experiments depicted in the illustration. In the example below, notice how the figure is further introduced with a figure citation, and a brief description of the procedure and samples. Not much more information, however, is provided about the cell proliferation assay. In general, samples do not need to be listed as explicitly as they are in this paragraph: you can instead introduce the samples as you describe the data.
- **Data description** highlights the trends or specific elements that you want the reader to notice and those aspects of the illustrations that support your conclusion. This means that the Results section is more than, "In order to determine X, we found Y (see Figure 4)."

You do not need to describe what happened with every sample at every condition—that's why illustrations exist! Describe the data well enough so that a reader does not need to look at the data. P values are especially useful to mention because they indicate the significance of the data. If your assay is qualitative (e.g., gel electrophoresis), try to estimate the differences by eye (or if possible by optical density measurements) instead of saying you have "more" product.

Although you will not have the space to describe all the data, you can acknowledge anomalous data to let the reader know that the data may not be perfect. Similarly, mention control samples to indicate how much the data can be trusted:

> Controls showed that colony formation depended upon the presence of DNA in the transformation mixture.

- The **conclusion** interprets the data, essentially answering the question implied at the beginning of the paragraph. But, you may ask, doesn't data interpretation belong in the Discussion? Yes and no. In the Results section, you only need enough data interpretation to help you provide context for the next set of experiments. In contrast, the Discussion section contains data interpretation in the broader scale (e.g. how it relates to the previously published literature).

Following is a sample from a different student paper that uses this organization (a corresponding paragraph from the Discussion section is found later in this chapter):

In order to explore the effect of the mutant p53 on the overall cellular response when a wild-type p53 allele is restored to a mutant cell, we compared the proliferation of two cell lines: one expressed the point mutant ($p53^{R270H}$); the other was unable to express

← Begin paragraph with rationale for experiment. Our sample paragraph states the rationale using a typical structure: "In order to

p53 (p53$^-$). In both cell lines, the allele on the homologous chromosome was p53^{LSL}, an inactive allele that we could restore to wild-type p53 activity by treating cells with OHT in ethanol (Figure 3). When no wild-type p53 was present (cells treated with ethanol only), both cell lines had similar doubling time of approximately 12 hours. These proliferation rates were kept relatively constant over the three passages. However, when we restored wild-type p53 activity (cells treated with OHT in ethanol), we saw a time-dependent decrease in proliferation in both cell lines. After twelve days there was a fivefold decrease in cell proliferation in the p53$^{R270H/LSL}$ cell line and a fifteen-fold decrease in cell proliferation in the p53$^{-/LSL}$ cell line compared to their ethanol-treated controls. Therefore, although both cell lines responded to the presence of wild-type p53 by slowing to a more normal rate of proliferation, cells expressing the mutant p53 plus the restored allele (p53$^{R270H/LSL}$) did not respond to the restored wild-type p53 as robustly as the cells (p53$^{-/LSL}$) which had only the activity of the restored allele.

> explore X, we performed Y." An alternative way to introduce the experiment could build upon the data of the previous figure: "Based on this result, we next set out to obtain Z information . . ."

> Introduce illustration with brief description of method and figure citation.

> Describe data in enough detail so that the reader does not need to look at illustration. Our sample includes such descriptive and quantitative phrases as "12 hours" and "fifteen-fold decrease."

> End paragraph with brief conclusion (e.g., answer to the question posed at beginning of paragraph).

Edit

There are several mistakes that we see so often in the Results section of student papers that we list them here.

Use the Correct Verb Tense

The Results section is largely written in the past tense because the data were obtained before you wrote the section (e.g., "The mice exhibited . . ."). Direct references to illustrations, however, should be in the present tense (e.g., "Table 1 lists . . .").

Avoid Overstating the Outcomes
Examples of overstatement are "The results clearly showed . . ." and "Figure 2 clearly shows . . ." Try instead to describe the data in an objective manner: "Cells expressing the Joh2 mutant SCC-13 migrated significantly faster than controls ($p = .05$) while SCC-4 and SCC-18 mutants migrated significantly slower than controls ($p = .001$)."

Avoid Saying That an Experiment "Worked" or Was "Successful" or That a Result Was "As Expected"
A true experiment does not have an expected answer: it may give results that are consistent with preliminary observations, or it may give completely unexpected information. In either case, if the experiment was carefully done, it has been "successful" and you should accurately report the results.

Be Precise in Using the Word "Significant" in Describing Differences Between Samples
Data are considered "statistically significant" only if they have been subjected to the appropriate statistical tests. The results of such a test can be found in Figure 2.11. The probability that the migration difference between a cell line and its control would be found by chance is less than 5 times in 100 tests is indicated by *, and less than 1 time in 1,000 tests is indicated by **. The result for SCC-2 cells could occur by chance more than 20 times in 100 tests, so is much less significant.

Sample Results
The subsections below are excerpted from a student paper that investigates whether a type 6 secretion system (T6SS) helps the bacterium *Helicobacter pylori* cause disease. The corresponding Methods for these subsections can be found at the end of the Methods section.

In Silico Analysis

As the first step in investigating *H. pylori* pathogenicity and whether it is influenced by a T6SS, we set out to map the T6SS component genes in the *H. pylori* genome to determine if *H. pylori* could have a functional T6SS. *In silico* analysis using protein BLAST comparisons is a powerful tool for finding functional homologs of target proteins. It has been successfully used to identify T6SS clusters in a wide variety of bacteria, many of which were later confirmed to be functional by *in vivo* and *in vitro* assays (5). *In silico* analysis of the *H. pylori* genome identified 14 conserved T6SS genes, belonging to 13 different Clusters of Orthologous Groups of proteins (COGs). These genes were found to be arranged in three distinct clusters, in supercontigs 4, 8, and 9 (Figure 1). All of the main structural components of T6SS were found to be present, as well as the known effectors Hcp1 and VrgG (Table 1). These results strongly suggest that *H. pylori* possesses a functional T6SS.

> Each subsection typically describes only one illustration.

> Remind reader of project goal.

> Explain purpose of experiment.

> Describe data well, and cite corresponding illustration.

> End description of experiment with a conclusion.

Growth and Motility

DnhB is an essential structural component of T6SS (22), so two deletion mutants of the DnhB gene were made and characterized to determine the function of T6SS in *H. pylori*. Growth dynamics of wild-type *H. pylori* and both mutants were determined in liquid media (Figure 2). No difference in growth was observed before 9 hours, however both Δ DnhB1 and Δ DnhB3 entered the stationary phase of growth by 35 hours, when the wild type still appeared to be in the lag phase. Wild type was in the log phase by 52 hours, and by 79 hours all three strains reached a similar level of growth. From these data we concluded that the T6SS affects the growth kinetics of *H. pylori* in liquid medium.

In many experimental systems, the T6SS plays a crucial role in motility as well as growth (9, 18). When motility of the mutant strains was compared to wild type on semisolid agar plates, a $57\pm8\%$ reduction in motility was observed in Δ DnhB1 and a $77\pm2\%$ reduction in Δ DnhB3 (Figure 3), suggesting that T6SS plays a role in *H. pylori* motility.

RESULTS CHECKLIST

Do

✓ Create a logical narrative based on the illustrations.
✓ Describe your data adequately so that a reader does not need to look at an illustration.
✓ Cite your data as an illustration or as "(data not shown)."
✓ Include rationale and conclusion for each experiment.
✓ Use the correct verb tense: Results generally uses the past tense because they were obtained in the past. However, you cite illustrations using the present tense (e.g., "Figure 2 shows . . .") because you are referring to data currently in the paper.

Don't

✓ Include information or data points that you will not describe in the text.
✓ Report irrelevant or inappropriate results.
✓ Overstate the results (e.g., "Figure 3 clearly shows . . .").

Methods: Document Your Process

The Materials and Methods section (hereafter, referred to as Methods) does not get much respect. In many biology journals, the Methods section is in smaller type and located at the end of the paper or available only online as Supplementary Material. These characteristics reflect the fact that most scientists do not read the Methods in the same way they read the rest of the paper.

Despite the lack of respect, the Methods section is essential because it describes the experimental design of the project, allowing biologists to replicate—and verify—each other's work. Biologists depend upon the reliability of other people's data in order to perform their own experiments. Thus, imitation may not only be the highest form of flattery but is essential for the field.

An equally important reason for the Methods section is that it helps readers evaluate the data. Although scientists skip or skim parts of the Methods with which they are familiar, they still want to be able to rapidly find the answer to any technical question that comes up when reading the Results section. Indeed, a reader's interpretation of the data may differ based on how animals were observed, or how data were treated statistically. The division of the Methods into subsections focused on a particular material or procedure of the project allows readers to quickly locate the information they seek.

Collect Your Content

The first thing to do when writing the Methods section is to gather details about the procedures you have used or are using. The Methods section includes five types of information: biological material, components, conditions, rationale, and data processing. Although not every procedure will have information for all of these categories, the categories are highlighted in the sample subsection below and then explained more in detail.

Western Blotting

Equal amounts of protein were loaded into each well of polyacrylamide gels. The following primary antibodies were used: anti-Shank1 (1:200, Neuromab), anti-Shank2 (1:100, Neuromab), anti-Shank3 (1:400, Santa Cruz), anti-PSD95 (1:1000, Thermo-scientific), anti-PanShank (1:1000, Neuro-mab), and anti-actin (1:3000, Sigma). IRDye 800CW and 680LT Secondary antibodies (Licor) were used at 1:5000 dilution for detection on an Odyssey IR laser Scanner (Licor). To provide a loading control, all anti-Shank and anti-PSD95 signals were normalized to the actin signal of the same sample. When neurons were infected with lentiviruses, data from infected neurons were compared to data from uninfected neurons within the same batch. Statistical significance was estimated with Student's t-test between infected and uninfected neuron cultures.

← Components, with sources

← Conditions

← Rationale

← Data processing

← Biological materials

Biological materials are descriptions of specific organisms or cell lines that are the subjects of the study. The description includes the species, source, genotype, growth conditions, and any ethics statement about how the material was treated. Genotype and sources could be listed in a table if several animal stocks or cell lines were used. Description of the organism is not necessary, however, if your experiments did not directly require cells (e.g., if *E. coli* was used only to overexpress the cloned mouse proteins that are the actual subjects of your study).

Components are the chemicals (reagents), kits, and equipment needed to perform the experiments. For components, the type of information that is needed is the source (company typically suffices because it can be difficult to determine the home city of multinational companies) and other information that may affect the outcome of a result, such as the catalog number of an antibody or the model number for a piece of equipment. Control samples could also be included, but one more often sees these samples in the legend, especially if the same method is used to generate the data for more than one illustration.

Conditions are the details under which a procedure was performed. This information allows a biologist to achieve the same results as those described in the paper. For example, were the animals observed in the early morning or midday? Were your tubes incubated at room temperature or 37°C?

Rationale for certain reagents or unusual steps helps a reader interpret the data. For example, ethidium bromide is typically used to visualize purified DNA in gels, but less well known is its ability to remove proteins that are bound to DNA. Therefore, you do not need to describe why you used ethidium in your DNA gels, but if you added ethidium when you lysed cells, you should explain that it was used to extract DNA-binding proteins during cell fractionation.

Data processing refers to information on how the results were analyzed. Examples include statistical analyses of numerical data, programs, and details of the microscope used to quantify fluorescent elements in a microscope image. Note that this section is used only if the processing is complex or unusual. For some experiments, standard tests, such as a X^2 test, can be mentioned when you describe results.

Strike the Right Level of Detail

The biggest challenge for students in writing the Methods section is to achieve a level of detail appropriate for the audience. In practice, you want to avoid rewriting the protocols or procedures you find in your laboratory manual. The Methods section is *not* a set of protocols (detailed list of steps to be carried out) but a collection of techniques that are described for *researchers who are already experienced with them*—even though the techniques may be new to you. The only details you should include are those necessary to replicate a result.

Due to the space limitations of a scientific research article, techniques that have already been described in a published research article do not need the same level of detail in your article. In such cases, you should cite the original article where the method was described and include any modifications that you made. We also recommend describing the general principle of the technique so that a reader can interpret your data without looking up the citation. Take a look at the sentences below.

Needs Improvement

To determine whether the mouse transgene was active in the fish cells, we measured the gene expression by the technique of Alpha and Beta (2013).

← Sentence does not indicate whether RNA or protein was measured.

Better

To determine whether the mouse transgene was active in the fish cells, we measured the gene expression by the modified RNA blot technique of Alpha and Beta (2013).

Saying that the amount of gene expression was assessed by a technique described in an earlier paper does not tell your reader anything about what the technique actually measures. Did you measure the amount of RNA or protein? A multitude of techniques can be used to quantify either, so knowing the way that expression was measured will help the reader interpret your data.

The best way to determine the appropriate level of details is to examine the relationship between protocol and methods. Following is a lab manual protocol for amplifying DNA, PCR. Compare it to the same content when it's rewritten as a Methods subsection.

Do not list samples or variables in Methods.

Add subheading to allow reader to find information easily.

Describe a single, complete reaction.

List reaction components (except DNA and enzyme) with final concentrations.

Protocol

1. Obtain 3 small sterile blue 0.5-mL PCR tubes. On the top and on the side of each tube, carefully label them #1 (template + primers), #2 (template only), or #3 (primers only). Also on the top of each tube put your bench number.
2. Set up the reactions as follows:

tube #1:	tube #2:	tube #3:
10 µL 5X Phusion® Buffer	10 µL 5X Phusion® Buffer	10 µL 5X Phusion® Buffer
5 µL 10X dNTP mix	5 µL 10X dNTP mix	5 µL 10X dNTP mix
5 µL 10X primer mix	–	5 µL 10X primer mix
5 µL template DNA (pET-Pfu)	5 µL template DNA (pET21b-Pful)	–
1 µL Phusion polymerase	1 µL Phusion polymerase	1 µL Phusion polymerase
24 µL ddH$_2$O	29 µL ddH$_2$O	29 µL ddH$_2$O

Methods

Polymerase chain reaction

Amplifications consisted of 50-µl reaction mixtures containing 1X Phusion® buffer, 1X oligonucleotide primer mix, 1X dNTP mix, 150ng template DNA (pET21b-Pful), and 2 units Phusion polymerase. Reaction samples were subjected to the following scheme: 95°C for 30 seconds in the first cycle, followed by twenty-five cycles of 95°C for 30 seconds, 55°C for 1 minute, and 72°C for 3 minutes, and a final 72°C for 10 minutes. The primers used were the forward primer [5' to 3'] TAATACGACTCACTATAGGG and the reverse primer GCTAGTTATTGCTCAGCGG.

Do not include details that would be familiar to someone who knows the technique.

3. Using a P20 set to 20 µL and a fresh tip for each tube, mix the reagents by pipetting up and down several times (try not to introduce any bubbles!). Spin all three tubes in the table-top microcentrifuge for ~20 seconds at 6000 rpm (2900 x g). Make sure to balance your tubes in the centrifuge! (When spinning the small 0.5-mL tubes, place them into an empty 1.5-mL Eppendorf tube as a holder before placing them in the rotor.)

4. Give your three tubes to your teaching assistant (TA). Your TA will place them in the thermocycler to run the following QuikChange program:
   1. 95ºC for 30 seconds
   2. 95ºC for 30 seconds
   3. 55ºC for 1 minute
   4. 72ºC for 3 minutes
   5. Repeat steps 2–4 twenty-four times
   6. 72ºC for 10 minutes
   7. 4ºC overnight

Protocol adapted from MIT 7.02 lab manual, Fall 2012

Notice the careful selection of details in the Methods section write-up. The only aspects described are the sequences of the PCR primers and the length and temperatures of the thermocycling protocol. A researcher experienced with PCR knows that these are the most important details for describing PCR. Given this information, your readers could probably use their favorite polymerase and obtain similar results. Your readers may even get a better yield of DNA.

Why are some details omitted in the Methods?

1. The Methods section describes techniques in a general manner. For example, the number and identity of samples are not specified. In addition, the volume and components of a *single, complete* reaction are listed. Such information allows a reader to scale up a reaction accordingly. Notice that most components of the reaction are listed with their final concentrations, rather than absolute amounts. The main exceptions to this rule are the template DNA and polymerase enzyme. The total mass of DNA is listed because DNA is usually quantified by optical density, which measures mass; enzymes are given in activity units because activity depends upon the number of active molecules, not the total number of enzyme molecules.

2. Your audience already knows them. That is, many details are unnecessary for an experienced researcher. For example, your reader will already know to use water to bring up a reaction to the desired volume and how to centrifuge a mixture to bring the reagents together.

Organize Your "Methods" so Your Reader can Easily Find Information

The structure of the Methods is characterized by the use of subheadings so that a reader can easily find information on your biological material or a specific procedure. This sequence is likely different than the order in which you performed the experiments in the lab. In general, the first subsections of your Methods should describe your main reagent, whether it is your biological material or RNA/protein. The remaining Methods subsections should parallel the experiments described in the Results.

A common mistake in structuring the Methods section is to follow the structure of the laboratory manual. Each of the techniques listed below theoretically could be described in a separate subsection:

Subcloning the gene
Mutagenesis of the gene
Purification of the plasmid for sequence analysis
Transformation of clone into an overexpressing strain
Overexpression of protein
Purification of protein

← Individual techniques that constitute a procedure should be grouped together, not described in separate subsections.

Instead, use subheadings to group similar content. For example, the techniques listed above could be combined into a single subsection with the subheading, "Mutagenesis, overexpression, and purification of protein."

Cell culture
RNA isolation and cDNA synthesis
Microarray analysis
Generation of stable overexpression and knockdown cell lines
Motility assay
Statistical analysis

← Describe main reagent and then organize to follow the order of your Results.

Although you may be tempted to write your Methods after the Results section, the Methods section could be written before instead because it just requires looking up information. Regardless of which section you write first, be sure to review the order of your subsections in both sections of your final draft.

Edit

Editing your Methods involves keeping an eye out for the following.

Use Clear Topic Sentences

Topic sentences are important in biology writing because they help orient a reader about the subject of a paragraph. Compare the two paragraphs below.

....

Needs Improvement

Samples were purified using the Qiagen® PCR purification kit, and digested with DpnI at 37°C for 16 hours to remove the methylated template DNA. The DNA was transformed into *E. coli* XL1-Blue for amplification, and cells were plated on LB-ampicillin agar plates. Two negative ("No DNA" and "Only template DNA") controls and a positive (pUC18) control were included. Pfu DNA was isolated from the transformants using the QIAprep Miniprep Kit (Qiagen®) and sequenced at the MGH Core DNA Sequencing Facility using the T7 promoter primer.

Paragraph lists a number of techniques: plasmid purification and digestion, transformation, plasmid isolation and sequencing. The topic is unclear.

Better

The PCR-mutated plasmid was purified and amplified for later sequencing. Samples were purified using the Qiagen®

PCR purification kit, and digested with DpnI at 37°C for 16 hours to remove the methylated template DNA. The amplified DNA was transformed into *E. coli* XL1-Blue for amplification, and cells were plated on LB-ampicillin agar plates. Two negative ("No DNA" and "Only template DNA") controls and a positive (pUC18) control were included. Pfu DNA was isolated from the transformants using the QIAprep Miniprep Kit (Qiagen®) and sequenced at the MGH Core DNA Sequencing Facility using the T7 promoter primer.

Added first sentence provides an overarching organization to the paragraph. Note how including the purpose of these procedures allows the sentence to refer to the methods without actually listing them.

Use Correct Verb Tense

The Methods section is written in the past tense because the experiments were performed in the past. Past tense is used for describing your procedures and your data analysis—for example:

> The Abel gene <u>was cloned</u> into the BamHI and SalI sites of the pSY vector.

This rule, however, does not apply when you describe the property of a reagent that is independent of you, as shown in this example:

> The Abel gene <u>was cloned</u> into the BamHI and SalI sites of the pSY vector. This vector <u>contains</u> that gene that confers resistance to the drug ampicillin.

In addition, actions in the Methods section are typically described using the passive voice to let the actions speak for themselves. One problem, however, with use of the passive voice is the introduction of dangling modifiers, which con-fuses who is actually performing the action. In the pair of sen-tences below, the second sentence has an additional word to clarify that the bacteria were not doing the washing.

Needs Improvement

After washing, the bacteria were spread on LB-amp plates.

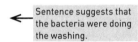
Sentence suggests that the bacteria were doing the washing.

Better

After being washed, the bacteria were spread on LB-amp plates.

Use Precise Language

Because the Methods section is used by scientists to replicate and interpret work, it is imperative to write precisely. To add precision to your writing, try these four tips:

- Describe the correct process. Beware of modifiers that could be interpreted to apply to all items of a list.

Needs Improvement

Cells were washed and fixed in 3.7% formaldehyde.

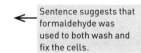
Sentence suggests that formaldehyde was used to both wash and fix the cells.

Better

Cells were washed with buffer and then fixed in 3.7% formaldehyde.

- Avoid the colloquial language used in laboratories. For example, one does not "run" a reaction or a gel. Instead, one "performs" a reaction; protein or DNA are "electrophoresed" or "separated by electrophoresis." In addition, centrifugation speeds are given in g, not rpm (as stated in your protocol), because g is independent of the rotor used (and rpm is not).

- Do not treat acronyms as nouns. For example, the phrase "PCR reaction" or "SDS-PAGE gel" make little sense when you remember what PCR and SDS-PAGE stand for polymerase chain reaction and SDS-polyacrylamide gel electrophoresis, respectively.

- Do not use a lab's own shortcut names for reagents or protocols. But how do you know the difference between a lab's shortcut name and the correct term? A good way to help identify the common terms is to read the Methods sections of papers from other laboratories. Special care should be taken for buffers because their composition can differ even within the same lab. For example, PBT is a common acronym for a buffer. The PB usually refers to <u>p</u>hosphate <u>b</u>uffered saline, the composition of which can differ for different organisms; the T could refer to Tween 20 or Triton-X100. Play it safe by listing the components of such a solution the first time it appears in the Methods, like so:

Drosophila cells were rinsed in PBS (130mM NaCl, 7mM Na_2HPO_4, 3mM NaH_3PO_4, pH 7.2) to remove traces of growth medium.
After fixation, cells were permeabilized in PBT (PBS plus 0.3% Triton-X100) for 10 min.

Sample Methods

The subsections below are excerpted from a student study on the potential role of a T6SS in the pathogenicity of the *H. pylori* bacterium. The corresponding Results for these subsections can be found at the end of the Results section.

....

Strains and Culture Methods

Helicobacter pylori (ATCC®43579TM) was propagated on commercial blood agar plates (Remel) or on blood agar plates supplemented with 25 µg/ml chloramphenicol to maintain the ΔIcmF mutations. This latter set of supplemented plates was made with Brucella agar (BD) and 5% defibrinated sheep blood (Quad Five). Bacterial cultures on the agar plates were incubated in a humidified 37°C incubator in GasPak jars filled with a microaerobic gas mixture (10% H_2, 10% CO_2, 80% N_2).

A stock culture of T84 human colonic epithelial cells (ATCC, CCL 6) was maintained in a commercial 1:1 mixture of Dulbecco's Modified Eagles Medium and Han's F12 Medi (DMEM:F12; ATCC), supplemented with fetal bovine serum (5% v/v), penicillin (100 U/ml) and streptomycin (100 mg/ml) according to ATCC's recommendation. Cell cultures were grown in 75 ml Falcon tissue culture flasks (BD) in a humidified 37°C incubator aerated with 5% CO_2.

In Silico Analysis

Sixteen conserved T6SS genes, as identified previously (5), belonging to 16 COGs (Cluster of Orthologous Groups of proteins, defined previously(15)) were used as bait to query the *H. pylori* genome using protein BLAST (pblast) searches via the Broad Institute website (broadinstitute.org) with standard parameters. All hits were mapped to the *H. pylori* genome to produce a map of all *H. pylori* T6SS proteins.

> First subsection describes the model organism.

> Explain the reason for nonintuitive steps.

> Detail components and conditions of procedures.

> The order of subsequent subsections roughly parallels the order of the corresponding subsections in the Results section.

Growth and Motility Assays

For the growth assay, 2-day-old *H. pylori* cultures were harvested from blood agar plates, centrifuged at 9000 g for 4 minutes, suspended in Brain Heart Infusion (BHI) medium, and diluted to an OD600 nm of 0.2 in rectangular 50-ml culture tubes (BD). Cultures were incubated under micro-aerobic conditions with continuous shaking at 100 rpm, and OD600 nm was measured at various time points.

For the motility assay, *H. pylori* cultures were harvested as described above. Each culture (2μl) was spotted onto motility plates containing Brucella broth (BD) and 0.3% (w/v) agar (BD). The plates were incubated for 24 hours under microaerobic conditions, and inspected for halos formed on the agar.

....

METHODS CHECKLIST

Do

✓ Provide methods for all the experiments you report in the Results section.

✓ Describe your components (biological material, other reagents and materials), conditions, rationale, and data processing.

✓ Provide enough detail so that a biologist experienced with the technique can replicate your result.

✓ Describe modifications you make to published procedures.

✓ Organize subsections and paragraphs in a logical manner: biological material first, then parallel with the Results.

✓ Write in a manner that is concise yet precise.

Don't

✓ Reiterate published procedures.

✓ List variables or number of reactions.

Don't

✓ Omit reasons for nonintuitive steps or reagents of a method.
✓ Write in strictly chronological order.
✓ Forget to use either subheadings or topic sentences.
✓ Use improper verb tense: Methods are generally written in the past tense.
✓ Include colloquial terms used in the laboratory (e.g., "run a gel," "PCR reaction").

Introduction: Start Broadly and Then Narrow

At a party, you might introduce yourself by answering a number of questions: What is your name? Where do you come from? How do you know the host (i.e., why are you here)? The Introduction of a research article answers the same sorts of questions:

- **Context**: What was known before the project?
- **Justification**: Why are you conducting this study?
- **Aim**: What is the goal of your project?

The Introduction is typically 500 to 750 words. The Introduction starts broadly with context (e.g., cancer, neurons) and gradually narrows to the justification and aim (e.g., to study a specific gene or protein).

While it may be tempting to start writing the context first, we recommend beginning with your aim and justification. Deciding upon your aim and justification will help you narrow and more clearly target the background context for your project.

Define Your Aim

The aim describes how you will address a gap in knowledge. Therefore, the aim not only states the goal of the project but may also point to the general methods that you used.

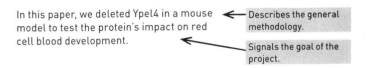

In this paper, we deleted Ypel4 in a mouse model to test the protein's impact on red cell blood development. ← Describes the general methodology.

← Signals the goal of the project.

Another type of aim could be your hypothesis, or a statement of what you expected. But what do you do if your experiment neither supported nor falsified your hypothesis? If you are still writing a paper about your research, the experiment must have shown you something unexpected and interesting:

> We hypothesized that deleting Ypel4 in a mouse model would lead to a defect in red cell blood development. The deletion, however, unexpectedly caused a stress response in all types of hematopoietic cells. Although the deletion was not suitable for testing our original hypothesis, we used it to investigate this unusual stress response.

These sentences show how your original hypothesis led you to do what biologists frequently do: manipulate a system in order to learn more about it. Such an approach will also help you develop a rich Discussion section later on.

Determine Your Justification

The justification explains why the project needs to be done. It is not enough to say that your lab instructor or lab supervisor told you to do so! There is a scientific motivation behind every project. Your job is to identify that motivation—even if it means asking your instructor or supervisor.

There are two types of justification:

1. The first type explains the importance of the project. The following statement explains reason why Ypel4 is important to study:

   > Although several of the Ypel family members are implicated in multiple functions central to the cell, the function of Ypel4 is still largely unknown.

2. The second type of justification is broader in scope and explains why the topic in general is being studied.

   > The final steps in differentiation of mammalian red blood cells are of particular interest because these steps give rise to the cells' unique phenotype and function.

Although both types of justifications are acceptable, you want to furnish the first type of justification (i.e., the motivation for *your* specific project). (Note that these justifications assume that your experiments either supported or refuted the original hypothesis. You would need different justifications if you switched to studying the stress response caused by Ypel4.)

Notice, too, the types of words that mark a justification statement: *although, however, but.* These terms emphasize the fact that the justification statement highlights the limit in our knowledge about a topic. Therefore, the Introduction describes not only what is known but also *what is not known*.

Highlighting what we do not know will help prevent you from making a common mistake: confusing justification with application or implication of the work. Your project may lead to a new cure for a disease or an improvement in environmental policy, but such applications did not prompt your specific research question. Your project builds upon past research, and addresses a gap in our knowledge. Therefore, justify your work by specifying that gap.

Develop Context for Your Study

The context is the background information of your project. Context for the Introduction section is a bit like a Literature Review in providing a survey of the field, but the Introduction should only include the research that is relevant to your project. Therefore, you probably do not need to start with proteins, DNA, or heaven forbid, "since the beginning of time." How do you decide how much is enough?

Determining the aim and justification of your project first will help you identify the topics that need to be explained in the context. Indeed, highlighting the key terms of our aim reveals questions that could be answered in the context: What is known about red cell blood development, and why are scientists interested in the process? What is known about *Ypel4*? If not much is known about this gene, what is known about genes related to it? You may also wish to comment on your methodology (here, a mouse model) in your Introduction, depending upon the novelty of your technique.

Start the Introduction with a Definition

A big challenge in writing the Introduction is crafting your first sentence. A simple definition of a biological phenomenon may serve the purpose because it lets your reader know the overall topic. Depending on the length of the definition, you also may be able to establish the overall significance of the topic. This opening sentence does both:

Erythropoiesis, or red blood cell develop- ⟵ Helps define the term
ment, is an essential process for any   erythropoiesis.
animal that has a circulatory system. ⟵ Helps establish signifi-
                                                 cance of topic.

Keep in mind that the specificity of your definition—or, indeed, the term you decide to define—will depend upon the audience of your article. Erythropoiesis may be fine for a general biology audience, but for a journal that specializes in red blood cell development, you can start with a more specific aspect of erythropoiesis.

Include the Minimum Amount of Information Necessary to Understand Your Project

Another challenge in writing the Introduction is avoiding irrelevant information. The relevancy of information can be determined by asking: Does the reader need to know this information in order to understand what I am doing, and why? Consider your audience, too: the context for a paper about the structure of a protein may differ if you want to submit the paper to a journal in the field of biochemistry as opposed to cancer.

The question about relevancy has an important second part: does the reader need to know this information *now*? Beginning writers frequently tend to insert everything they know about, say a protein, the first time it is mentioned. The reader, however, may not understand the relevance of some of the information until later in the paper, perhaps after seeing your Results section. In that case, fear not: you will find that this information fits easily into your Discussion section.

Let us say, for example, that you devised a clever genetic screen in Drosophila to detect fly proteins that might help biologists understand human muscular dystrophy. You identified protein X through this screen, then went on to discover that the sequence of protein X had some commonalities with sequences of proteins found in humans, fish, and birds. In the Introduction, you should explain what is unknown about this human disease and then justify building a fly model of the human disease. You should not include information about protein X's similarity to proteins in other organisms in the

Introduction; doing so would divert from the argument for using flies to study the human disease. Worse yet, the information does not give the reader much new information because a large majority of proteins share motifs with other proteins. The information would be more appropriate in the Discussion, where you can explore what the shared motif tells you about the function of your new protein. In this case, if you showed that protein X was expressed in fly leg muscles and one of the shared motifs was with a protein that bound the cytoskeleton, then you can suggest in the Discussion that protein X was involved in strengthening muscle cells.

Structure your Introduction

Once you have determined the aim, justification, and context of your Introduction, turn your attention to the structure. The Introduction is shaped like a funnel: the scope of the beginning is broader at the beginning than it is at the end. This structure makes sense from a reader's point of view: a reader is more likely to understand your work if you can contextualize your work within the field.

A simple way of achieving a funnel-like structure (Figure 2.12) is to arrange your paragraph topics in decreasing scope: context, justification, aim. The context helps orient readers to the topic

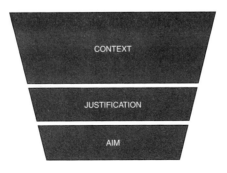

FIGURE 2.12 Structure and Content of an Introduction.

of your research. The justification—highlighted by words like "however" or "little is known"—would be situated right before the aim because it highlights the gaps in our knowledge. Your aim should be at the end because it is the most narrow in scope, and it provides a smooth transition to the Methods and Results sections.

One last thing: although Figure 2.12 situates the aim at the end of the Introduction, some scientists end their Introductions with a summary of the project. The advantage of this arrangement is that the reader knows what to expect. The disadvantage is that the summary is already present in the abstract, which appears right before the Introduction. So, if you prefer to summarize your key results in the Introduction, be sure to use different words and phrasings.

Edit

After the title and abstract, the Introduction may be the first section that your reader reads. So, polishing your language helps make a good first impression.

Use Correct Verb Tense

The Introduction is typically written in the present tense except when you are referring to specific people. Take a look at these two sentences:

The structure of DNA was determined by Watson and Crick.

DNA is a double helix.

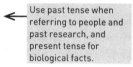
Use past tense when referring to people and past research, and present tense for biological facts.

The past tense is used in the first sentence because Watson and Crick conducted the study in the past. Their conclusion conveyed by the second sentence is expressed in the present tense because it states a biological fact.

Use the First Person Plural When Describing Your Own Work

Biologists usually use first person plural (e.g. "We have shown that . . .") even when they refer to a single-author paper. This is done because most projects have help from many people (e.g., labmates, technicians) whose contributions are important but not large enough to merit authorship.

Format Your Citations

The Introduction provides the intellectual background for your project, so this section will probably have the most citations to the literature. We describe some formats for citations in Chapter 7, but it is important that you use the citation format required by the publication to which you will submit your research article.

Sample Introduction

Our sample Introduction is reproduced from a student paper on the role of Ypel4 in red cell blood development. It was developed using the same information we gave as examples for context, focus, and justification.

Erythropoiesis, or red blood cell development, is an essential process for any animal that has a circulatory system. In humans, approximately 2 million reticulocytes, the final precursor to mature red cells, are produced in the bone marrow every second, and production can increase up to 20-fold under severe hypoxic stress (An & Mohandas, 2011). Through the course of their development, mammalian

> First paragraph begins context by defining the broad topic (red blood cell development) and explaining the significance of the process.

erythrocytes lose their nuclei and undergo a substantial shape change before becoming functional erythrocytes circulating in the bloodstream.

The final steps in differentiation of mammalian red cells give rise to the unique phenotype and function of erythrocytes. Terminal erythropoiesis is a complex yet very well coordinated process that involves many cellular changes, including decrease in cell size, change in cell shape, increase in chromatin condensation and eventual enucleation, increase in hemoglobin synthesis, and changes in cell surface receptor expression (Hattagandi et al., 2010). Mechanistically, the signals and processes that enable erythroblasts to undergo this transformation are just being understood.

Second paragraph focuses on a specific stage of red blood cell development: terminal erythropoiesis.

We recently identified *Ypel4* (*Yippee-like 4*) as a gene of interest based on its high induction during terminal differentiation of murine erythroid cells (Wong et al., 2011). The *Ypel4* gene has been virtually unstudied since its discovery in 2004 as part of the highly conserved *Ypel* gene family, which is present in diverse eukaryotic species from fungi to humans (Hosono et al., 2004). The *Ypel* genes bear similar sequences to the Drosophila gene *Yippee*, an intracellular protein that contains a putative zinc-finger-like protein-binding domain (Roxström-Lindquist and Faye, 2001). Limited studies have been done on *Ypel* family members, but have implied functions central to the maintenance of life on a cellular level, including adhesion, cytoskeletal distribution, cell cycle regulation, growth inhibition and senescence, and cytokinesis (Farlie et al., 2001; Baker, 2003; Hosono et al., 2010). Previous experiments have suggested that all of the *Ypel* family proteins, except for YPEL4, are ubiquitously expressed in all tissues in mice and humans (Hosono et al.,

Third paragraph explains how Ypel4 became a candidate involved in terminal erythropoiesis.

Introduction briefly shifts away from context by stating the justification (i.e., what is unknown about Ypel4), and ends with aim of the study.

2004). Here, we probe the potential function of Ypel4 in terminal erythropoiesis by examining Ypel4 expression and loss of function in mouse erythroid cells.

INTRODUCTION CHECKLIST

Do

✓ Formulate an aim that is clear and allows you to learn something regardless of the outcome.

✓ Justify your work by identifying the gap in knowledge that your research addresses.

✓ Develop context that is relevant to your goal and appropriate for your audience.

✓ Create a funnel-like overall structure: start broadly and end with your goal.

✓ Define terms that are unfamiliar to your audience.

✓ Use past tense for past actions and present tense for biological principles.

Don't

✓ Include unnecessary background or repetition in your context.

✓ Exaggerate (or understate) the importance of your work.

✓ Describe new results in the same way you describe them in the abstract.

Discussion: Interpret Your Data

The purpose of the Discussion section is to make an argument, not simply explain everything that went wrong with your experiment. But what kind of argument? In the Discussion, biologists argue the significance of their work by comparing their data to the published findings of other scientists in the field and considering the broader implications and limitations of their work. In other words, they don't just explain; they interpret.

Your Work is Important!

Sometimes it's hard for students to make an argument about the significance of their research because of its preliminary nature. We perfectly understand that most work performed in college is not ready for publication, but your work is important because it represents the beginning of your research career as a biologist. So what can you do?

You can still discuss how your work would contribute to the field when (not if!) it is completed. Be sure to back up your claims by citing other people's work. You can also discuss the limitations of your approach *and* explore alternative experimental approaches. Such discussion may help give you ideas on how to resume your research, or may prove valuable to another student who continues your project. In the end, the Discussion section should give you an opportunity to think more deeply about your project and your future contributions to biology.

Formulate Your Argument

The first thing you should consider is how well your work addresses the aim of your Introduction. Perhaps you didn't answer the question implied by the aim or answered a completely different question. If either case applies, change your aim—it's much easier (and more ethical) than changing the data! You'll find yourself in good company, too. It is not at all unusual for a scientist to find the experiment has answered a different question from the one asked initially—and this new answer can be important.

Here is an example of an aim and the corresponding answer:

Question: How do metabolites A, B, C affect the progression of pancreatic ductal adenocarcinoma (PDAC), a form of pancreatic cancer?

> Answer: Metabolites A, B, C affect PDAC progression via muscle cells, and not via diabetes.

The answer to your question (whether or not it has changed) is your paper's overall argument, which anchors your Discussion. Once you have identified your argument, select the illustrations for your results that provide the crucial evidence for your argument (you may cite these illustrations in the Discussion as you expand your argument).

Another good strategy is to design a diagram that summarizes the findings of your work in the context of what's already known. Figure 2.13 uses the key terms of the student's conclusion above to connect her finding with the known literature.

Identify Your Contributions to the Literature

In the Introduction, you explained the scientific basis of your research (i.e., how previous work led up to the goal of your

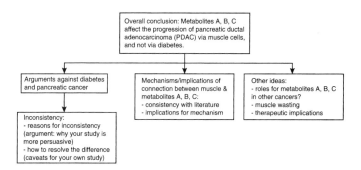

FIGURE 2.13 Brainstorming Ideas for the Discussion Section. Note how key terms in the conclusion are connected to the literature.

project). Similarly, your Discussion compares your work to those in the literature to highlight your contributions to the field. Figure 2.14 has a flowchart of questions that may help you identify your place in the dialogue of science.

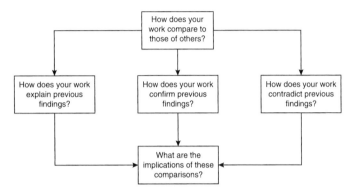

FIGURE 2.14 Ways to Compare Your Work to Those of Others.

The questions are explained in more detail below.

Explain Previous Findings
One possible outcome of your study is that it helps explain observations of others. In the example here, a student's analysis of mutant enzymes sheds light on why cancer cells preferentially express these mutations.

The lowered activity and allosteric activation of the mutant PKF4 enzymes found in human cancers support the hypothesis that cancer cells preferentially express PKF4 because of the ability for its activity to be down-regulated, unlike the PKF3 isoform. Because these mutations were found in cancers of human patients, they may provide cancer cells with some sort of selective advantage. We therefore suggest that these mutations are advantageous for the proliferation of cancer cells because they lower pyruvate kinase activity, which is important for the metabolic program of a proliferating cell (6).

← Citation of data from the student paper.

← Previous observation to be explained by data.

Confirm Previous Findings

Another possibility is that your work confirms other people's data or is consistent with them. In the paragraph below, a student compares her mutation, H147K, in the *Pfu* DNA polymerase with published data on the KOD DNA polymerase. This paragraph ends with an implication.

Our results with Pfu H147K are consistent ← Highlight consistency, with those of Kuroita et al. (2005), who found and cite supporting that positive residues at the active site in literature. KOD were best suited to interact with the thumb domain and thus to give the enzyme a higher fidelity. Lysine is more positively charged than histidine, and the resulting mutant Pfu displayed a lower mutation rate than the wild-type protein (Figure 4). The similarity in behavior between our work with Pfu and the work in KOD by Kuroita et al. (2005) suggests that fidelity in the ← Explicitly state implica- B family of DNA polymerases depends tion of consistency. upon the charge interaction between the active site and the thumb domain.

Contradict Previous Findings

It is also possible for your work to contradict the work of others. Don't despair, because this option opens up a wealth of points you can consider in your Discussion. What could account for the observed differences? What experiments could be done to resolve the differences? Don't assume that your data resulted from experimental error! Assuming your data are correct, what else could account for the differences? Consider the methods used to generate the data—are there differences in sensitivity? Do they measure different aspects of a problem? Another possibility is that the contradiction reveals something deeper about the biology.

In the following example, notice how the student suggests explanations for her finding that her mutant p53 proteins partially inhibited a wild-type protein, whereas previous scientists reported that a different mutant p53 protein completely blocked wild-type activity. (The corresponding paragraph from the Results section of this paper is found earlier in this chapter.)

Restoration of a wild-type p53 allele was sufficient to induce cell cycle arrest in our sarcoma cell lines, even in the presence of a mutant p53 protein. This observation is not consistent with previous research (Brosh and Rothar, 2009), which suggests that the mutant form of p53 has a dominant negative effect over the wild-type protein. This discrepancy could be due to the nature of the mutations we are working with; it is possible that these mutants (p53R270H and p53R172H) are not potent repressors of the wild-type p53. It could also be due to the presence of multiple oncogenic stimuli in our sarcoma cells if these effects overcome the inhibition by the mutant p53 and lead to activation of the wild-type protein.

← Signals a comparison with the published literature.

← Explains some reasons for the discrepancy. Suggests that the discrepancy may reveal unexpected complexity in the interaction of different mutant proteins with the wild-type proteins.

Consider the Implications

Regardless of whether your work explains, confirms, or contradicts the work of others, explore the implications of your findings. Essentially, you want to state the biological principles that your work helps to build. As you revisit the literature while developing the Discussion section, broaden the scope of your inquiry. Articles that were irrelevant for the Introduction could be appropriate for the Discussion. For example, let us say you are studying a protein involved in worm development. What are the implications for our understanding of worm development? If the protein has a homologue in humans, you could also consider the implications for human development. What are the evolutionary implications? Finally, consider the type of protein you are studying. Does this protein share amino acid sequence motifs with other proteins involved in other systems (e.g., nervous system, immune system)? If so, does your work suggest that that the protein acts in the nervous system? the immune system?

Admit Faults

(Un)fortunately, science does not end with the writing of a paper. A scientist understands that no matter how thoroughly one performed a study, there are always unanswered questions or assumptions that require further validation. Therefore, your Discussion should acknowledge the limitations of your work before your reviewers do.

Limitations are not experimental error (e.g., you pipetted the incorrect amount of reagents or forgot to feed the animals). Think, instead, about the limitations of your interpretations and of your methods. How confident are you in the precision of your measurements? What could be done to make the data stronger? What alternative interpretations can be made about the data, and how could you test them?

In the following example, notice how limitations can give you an opportunity to suggest new experiments. Your work may suggest a novel function of a gene; the function may be outside the scope of your study but would be an excellent starting point for a future project. Such suggestions for experiments expand your contribution to science.

> Our in vivo morphological evidence supports in vitro observations, but does not explain how the decrease in neuron cell size relates to the formation of memory. Determining whether the relation we have seen is either causal or correlative will be an important goal in future research.

Structure Your Discussion

The Discussion section may be the most difficult section to write because it does not have a clear structure like the other sections of a scientific research article. A common mistake is to reiterate the Results section with data interpretation sprinkled throughout.

Although the body of the Discussion does not seem to have an obvious overall structure, it may help to remember the hourglass metaphor for a research article (see Figure 2.2): the Introduction starts broadly and narrows, while the Discussion section begins narrowly and widens out in scope.

Your opening paragraph has a narrow focus because it reminds the reader of your project. Your second paragraph could extend the scope of the Discussion by comparing your study to those articles that directly lead to your research question. Subsequent paragraphs could then elaborate on broader aspects such as your data's implications for the biology of your model organism or other proteins that are homologous to your protein of interest.

Develop the Overall Structure by Considering Your First and Last Paragraphs, and Selecting Your Key Experiments

Start your Discussion by composing your opening and closing paragraphs to help focus your argument. Start the first paragraph by reminding the reader of what you set out to investigate and why. Such a paragraph allows the Discussion to be read independently of other sections but also connects the Discussion with the Introduction (helpful for readers who skim your article). A good ending for the opening Discussion paragraph would be your overall argument.

The closing paragraph of your Discussion is your conclusion, and it can be written in as little as two sentences. The first sentence of the closing paragraph reiterates your overall argument.

YPEL4 is a member of a highly conserved family of putative zinc-finger-containing proteins present in diverse eukaryotic organisms, from slime molds to humans (Roxström-Lindquist & Faye 2001). This evolutionary conservation suggests important function on a basic developmental or cellular level. The functions and molecular mechanisms of the YPEL proteins require further elucidation, as their role in eukaryotic life is still not well understood. In this paper, we identified YPEL4 as an erythroid-specific protein that plays a role in the terminal differentiation of red blood cells.

← Opening paragraph reminds reader of the general topic.

← Reminds reader of justification of project.

← States the overall argument of the Discussion.

The second sentence then highlights a broad implication of your work. No matter how short this paragraph is, remember it is your final word. So, consider it carefully.

In summary, we identified Obc as a candidate gene that suppresses motility in cancer cells. Our work contributes towards a greater understanding of metastasis, and draws us closer to the goal of mediating therapeutic intervention for this complex process.

← Last paragraph reminds reader of overall argument, and points towards the future.

Remember, too, that your Discussion is an argument that expands the contributions of your work to biology. Therefore, concentrate on experiments that contribute most to your overall argument. Doing so would prevent you from reiterating the overall structure of the Results section. For example, let us say you wanted to examine the role of a particular amino acid in the function of an exonuclease: you mutagenized the gene, expressed and purified the protein, and assayed the mutant protein for DNA binding and exonuclease activity; the mutant protein gave conclusive data for the exonuclease activity assay but not DNA binding. Your Discussion should focus primarily on the activity assays because they are most relevant for understanding exonuclease function. In addition, exonuclease activity should be discussed before DNA binding (even if the order was switched in the Results section) because the DNA binding assay returned inconclusive data.

Structure Your Paragraphs to be Suitable for an Argument
Unlike a paragraph in the Results section, an effective Discussion paragraph is structured like an argument: begin with a claim, then follow it up with evidence and analysis. Let us see how these elements work together. In the paragraph here, a student compares her Pfu mutation with a KOD mutation described in the literature; the comparison allows the student to propose experiments that could help understand the role of an amino acid (I142) in polymerization activity.

....

The I142 residue plays a role in the polymerization activity of the B family of DNA polymerases. Just as Kuroita et al. (2005) saw a change in activity when the KOD DNA polymerase was mutated at I142, so we saw a similar change in activity in Pfu mutated at I142. Interestingly, whereas our Pfu I142R showed lower polymerization activity (Fig. 4), the KOD I142K protein displayed higher polymerization activity compared to wild-type protein. The differences in the polymerization activity may result from structural variations elsewhere in the exonuclease regions of KOD1 and Pfu, or between lysine and arginine, which alter how a residue at 142 interacts with the

← States as a claim a biological principle based on your data.

← Signals evidence to support claim. Corresponding figure in the Results is also cited.

← Offers reasons and implications for inconsistency with previous research.

rest of its respective polymerase. Since our
assay allows only qualitative comparisons
between proteins, determination of elonga-
tion rates of the mutant DNA polymerases
would allow us to compare more directly the
activities of I142R and I142K, and more fully
understand the function of I142 in polymerization activity.

Highlights a limitation
of study, but suggests
how to extend the study.

This paragraph starts out with a claim. A claim could simply be a conclusion for one of your illustrations, or it could be a statement that compares your work to others (e.g., "Our work confirms/contradicts . . ."). Note the scope of the claim: the claim states biological principle (e.g., role of an amino acid in protein activity) rather than an observation of the data (e.g., decreased activity with a mutation). Making such claims at a higher level allows you to compare your work more easily with those of others.

The paragraph continues with evidence, which can consist of your data (you could cite certain illustrations) or the data of others. Not every piece of data needs to be highlighted— just the crucial experiments—and the data do not need to appear in the same order as in the Results. Most importantly, the data are not described with the same level of detail as in the Results section: a comparison is made between the muta- tion and wild-type proteins, but the reader does not need to be given more information about the experiment. For this reason, we recommend citing the figure or table number so that a reader can easily find the experiment in the Results section.

The paragraph ends with analysis that explains the logic connecting the evidence to the claim. You might also intro- duce a limitation of your project, or comment on what could be done to strengthen your claims. Integrating limitations throughout the Discussion section is more elegant than lumping them together in a paragraph.

Edit

The Discussion is the section where you are allowed to speculate about your data, giving your reader the benefit of the many things you have thought about while conducting your experiments. As with other sections, there are some details that need special polishing in this section.

Use Correct Verb Tense

The Discussion section is written largely in the present tense. In the Results section, you discussed your findings in the past tense because you knew they were accurate for the experiments that you had completed. Now you are assuming that the biological principles you uncovered in those experiments are generally true and would hold up no matter how many times you repeated the experiment. Therefore, you can discuss these principles as facts, just like you describe scientific facts in the Introduction section.

Select the Right Verbs to Achieve the Right Level of Speculation

Avoid overspeculating: cite your evidence and explain your logic. Avoid underspeculating: give yourself some credit. You may think that six weeks on a project did not produce much data, but your work counts!

Choosing the right verb will help you speculate at the appropriate level. For example, the verbs "indicate" and "suggest" both mean "to show, demonstrate." The word "indicate," however, is stronger, so it should be reserved for times when the data are very persuasive. In addition, you use the word "prove" for the first time something is shown to be true, and "confirm" for subsequent times.

The amino acid sequence of the mutant enzyme differed from that of the wild-type protein at two positions. One of the differences, leucine at position 255, replaced a proline that is conserved in this enzyme from many other species, **suggesting** that loss of this proline is responsible for the loss of enzyme activity. When the mutant gene was genetically engineered to again encode proline at position 255, the expressed enzyme was fully active, **indicating** that this proline is required for enzyme activity.

> "Indicate" is a stronger term than "suggest," so use it only when your data are especially persuasive.

Avoid Excessive Hedging

Hedging helps you avoid overstating the importance of your data. However, you still want to convey some level of confidence in your work. Therefore, limit your sentences to no more than one "hedge word" per sentence. As you can see in Table 2.5, hedge words can appear as verbs, nouns, or adjectives.

TABLE 2.5 **Common Hedge Words**

Nouns	Verbs	Adverbs
appearance	appear	apparently
possibility	may be	possibly
probability	seem	probably
suggestion	suggest	seemingly

Sample Discussion

The sample Discussion is excerpted from a student project on the role of TrkB in PSD-95 expression in mouse neurons.

The numerous molecular components affecting synaptic plasticity at dendritic spines have been well identified in the past several decades; however, these compounds have yet to be linked together into a comprehensive picture. Recently, Yoshii and Constantine-Paton (2007) compiled a coherent pathway: NMDAR activity, BDNF-TrkB activation, and PI3K/Akt pathway act together in a positive feedback loop to deliver PSD-95 to the excitatory dendritic puncta synapses in the visual cortex and alter their plasticity. Their in vitro study provides a more comprehensive picture of the interconnected mechanisms that affect synaptic plasticity, but is not a sufficient model for physiological conditions. Our in vivo study and detailed morphological analysis of PSD-95 expressing neurons provide a valuable perspective to this important regulatory pathway. Specifically, we show that TrkB contributes to PSD-95 synthesis in the cell soma and to overall dendritic maturation.

◄— Remind reader of justification of project.

◄— End first paragraph with overall conclusion.

Inhibition of TrkB by 1NM-PP1 led to a decrease in the cell somas of PSD-95 expressing neurons (Figure 2). This decrease in cell soma size can be explained by altered activity of other components downstream of TrkB activation in the PI3K/Akt pathway, such as mTOR (Kwon et al., 2003; Kumar et al., 2005) or the Golgi transport mechanisms needed for efficient PSD-95 delivery (Yoshii and Constantine-Paton, 2007). Alternatively, the smaller cell soma size could be due to constitutive activation of phosphatase and tensin homolog (PTEN), a tumor suppressor gene that has been shown to inhibit the PI3K/Akt pathway and potentially alter synaptic plasticity (Kwon et al., 2006). Although our proposed

◄— Use literature to help explain your findings.

mechanisms are consistent with previous observations, it is still unclear what a change in cell soma size indicates about synaptic plasticity (Xu et al., 2000).

> Highlight a future direction of the research.

Our work is also consistent with studies that demonstrated that BDNF-TrkB pathway regulates dendritic growth and maturation, specifically in distinct layers of the visual cortex as well as in different types of dendrites. (Kumar et al., 2005; McAllister, Lo, and Katz, 1995). PSD-95 expressing neurons with inhibited TrkB function also show visibly less dendrite maturation after eye opening, particularly in basal dendrite arborization (Figure 3B). Basal dendrites showed a more varied response than apical dendrites to BDNF across different layers of visual cortex, suggesting that factors besides BDNF and TrkB activation affect the arborization and synapse formation across the visual cortical layers. Characterizing this difference in arborization in future studies may provide a more complete picture of synaptogenesis and how PSD-95 is trafficked into the synapses to induce synaptic plasticity.

> Compare your findings to those of the published literature.

Our study demonstrates that synaptic plasticity is indeed regulated by the NMDAR-BDNF-TrkB-PSD95-PI3K/Akt positive feedback loop pathway in vivo. Our morphological analyses of these PSD-95 visual cortical neurons suggest that TrkB inhibition can induce drastic structural changes at the synaptic level, particularly at the level of dendritic synapses. Thus, the molecular changes underlying these structurally important changes are likely to have implications for synaptic plasticity and mechanisms of efficient synaptic transmission.

> End Discussion by reiterating the conclusion and implications of the study.

DISCUSSION CHECKLIST

Do

✓ Explain your contribution to the field by comparing your work with those of others.

✓ Include limitations of your methods and interpretations.

✓ Provide evidence and analysis for each claim that you make.

✓ Introduce your Discussion with a paragraph that reminds the reader of what you did and why, and end with your overall conclusion.

✓ Focus on the question set up in the Introduction or the data in the Results.

✓ End with a paragraph that summarizes the project's overall conclusion and impact.

✓ Speculate at the appropriate level, not too much or too little.

✓ Use appropriate verb tense.

Don't

✓ Regurgitate the Results section: instead, elaborate only on the key data, and avoid redescribing the data.

✓ Write arguments that are wordy, unclear, meandering, and so forth.

✓ Repeat the structure of the Results section: write an argument, not your story plus interpretation.

✓ Hedge excessively: use no more than one hedge word per sentence.

Title: Get Noticed

You can think of the title as the brand name, or label, of your research article. Scientists tend to select the papers that they read by rapidly scanning long lists of titles, so the title is the first, and frequently the only, chance your paper has to catch a reader's attention.

An effective title:

- Accurately reflects the content and purpose of the research article
- Provides key words for databases
- Is understandable to a wide variety of biologists

Clearly titles have their work cut out for them.

Your Title is an Argument

Constructing an effective title involves converting your overall argument into a concise, informative phrase.

While developing a title for publication in a journal, you should consult the Instructions to Authors for that journal's specific criteria regarding the number of characters, whether title words are capitalized or in sentence case, and so forth. We will discuss more general goals for a title: concision and comprehensibility to a broad audience of biologists.

How do you go about constructing a title? Compose a title only after you have completed the final version of the paper. Let us say that you have confirmed a new pathway (a set of interacting proteins) involved in the formation of memories. So, a basic title could be something like, "Studies of Memory Formation," but it is too generic and does not sufficiently describe your work.

Convert your generic title into a title that is specific, concise, and not filled with jargon.

State Your Conclusion Using Two to Four Key Terms

One of the purposes of a title is to provide key terms. Such terms allow your reader to find your article easily in an online database. Good key terms would be the organism and/or proteins you studied, and perhaps the way in which the work was done.

Let us say that the pathway you studied involves three proteins found in the brain: brain-derived neurotrophic factor (BDNF), neuronal tyrosine kinase B (TrkB), and postsynaptic density 95 (PSD-95). This pathway was initially identified *in vitro*, but your experiments demonstrated that the pathway is active in *vivo*. You can include all of these terms in your title by conveying the overall argument or conclusion of your work:

> An In vivo Demonstration of Brain-Derived Neurotrophic Factor (BDNF), Neuronal Tyrosine Kinase B (TrkB), and Postsynaptic Density 95 (PSD-95) Working Together in the Formation of Memories

This title incorporates all of the key terms, but it is still not very good, right? It is wordy and exceeds the maximum length of a title allowed by many journals.

Be Concise

There are a number of ways to condense your title. You can cut extra words (e.g., "an," "working together"):

> In Vivo Interaction of Brain-derived Neurotrophic Factor (BDNF), Neuronal Tyrosine Kinase B (TrkB), and Postsynaptic Density 95 (PSD-95) in Memory Formation

> Consider a colon:

> Interaction of Brain-Derived Neurotrophic Factor (BDNF), Neuronal Tyrosine Kinase B (TrkB), and Postsynaptic Density 95 (PSD-95) in Memory Formation: An In Vivo Demonstration

Use an active verb:

Brain-Derived Neurotrophic Factor (BDNF), Neuronal
Tyrosine Kinase B (TrkB), and Postsynaptic Density 95
(PSD-95) Work Together to Form Memories In Vivo

You would be fine in choosing any of these titles, but we can
make them a bit better.

Use Abbreviations and Technical Terms Judiciously

You want your title to be as accessible as possible, so abbreviations should be avoided, unless they are commonly understood
abbreviations like DNA, RNA, and ATP. Names of genes and
proteins also do not need to be spelled out if you think your
audience (depending on the field and journal) does not need it
or if the full name does not provide relevant information (most
proteins are given a name before their real function is known).

Applying this advice can be tricky, but it works well for our
sample title. The title already has terms (e.g., proteins, memory
formation) that tell readers outside the field enough to decide
whether they want to read the paper; the complete names of
the proteins give only slightly more information than the abbreviations. Using the abbreviated names of the proteins considerably decreases our sample title's length:

In Vivo Interaction of Proteins BDNF, TrkB, and PSD-95
in Memory Formation

Another solution is to avoid listing the proteins altogether,
and focus instead on the confirmation of the new pathway in
which these proteins are involved, like so:

In Vivo Demonstration of a New Pathway for Memory
Formation

These last two titles are both good but aimed at different audiences. You might want to use the one listing the proteins for a paper published in the primary literature. The title focusing on the pathway might be better for a literature review or a talk to a general biology audience.

In both cases, we kept the key term "memory formation" but we were much more specific about the kind of study that we conducted—*in vivo*—and signaled to readers that we were investigating protein pathways. Those three elements likely mean that anyone searching a database for studies on proteins and memory formation would be likely to find our article. But before we finalize our title, let's do one more thing: check our grammar.

Make Sure That Prepositional Phrases are Modifying the Correct Terms

Let us compare two versions:

Needs Improvement

Confirmation of a New Pathway for Memory Formation Using In Vivo Studies

Better

In Vivo Confirmation of a New Pathway for Memory Formation

The first title suggests that the formation of memories (rather than the author) is using *in vivo* studies because the preposition "using" follows the term "formation." The second title shows that *in vivo* refers to confirmation.

TITLE CHECKLIST

Do

✓ Compress your overall argument into a concise, informative phrase.

✓ Include terms that are understandable to a broad audience of biologists.

✓ Adhere to the journal format for length and structure.

✓ Write concisely: every word counts.

✓ Use abbreviations and technical terms judiciously.

Don't

✓ Be vague about the content of your work.

✓ Include jargon.

✓ Write imprecisely: make sure that phrases are modifying the correct antecedent.

✓ Be wordy.

Abstract: Advertise Your Work

Along with the title, scientists use the abstract to determine the relevancy of articles retrieved from a database search. The abstract is your advertisement pitch: it is a concise, independent summary of your research paper. Let's take that phrase apart:

- An abstract should be **concise**. Most journals have word limits of 150 to 250 words, so every word must count.
- An abstract should **stand alone** or be "independent" because abstracts are frequently placed in databases and conference proceedings—in other words, places that do not have room for the full research article. Therefore, an abstract should not cite other research articles or illustrations from the article.
- The abstract **summarizes** your research article and announces or implies its purpose. Technically speaking,

this means that material from all sections of a scientific research article—Introduction, Methods, Results, and Discussion—should be in your abstract. However, the sections are not represented equally in a scientific abstract.

Write Your Abstract Last!

Write your abstract *after* you have written the rest of your paper. Doing so will enable you to identify the key information from the article and distill it into a few sentences.

Structure Your Abstract

In general, abstracts tend to have the following structure:

- **Introduction**: one or two sentences that provide context but more importantly convey the aim or goal of the work. The beginning of the abstract should be especially informative because it is the first thing the reader sees.
- **Methods**: may not be stated explicitly. At most, the overall strategy may be stated in half of a sentence; other procedures can be mentioned as the Results are described (e.g., "Immunoblots show . . .").
- **Results**: two or three sentences that describe the key findings of the work. By "key," we mean the data that provide the strongest support for the research article's conclusion. The description should be as specific as possible, so consider including numbers.
- **Discussion**: typically just one sentence that conveys the overall conclusion and impact of the research. The conclusion should resonate with the aim stated at the beginning of the abstract, as well as the conclusion stated at the end of the Discussion section. Make sure that the abstract and Discussion section come to the same conclusion (without using the same words)!

Let's see how this structure is revealed in a sample:

Terminal erythropoiesis, or red blood cell development, is an essential but very well-coordinated process that is only beginning to be understood. To elucidate the steps of terminal erythropoiesis, we evaluated the role of Ypel4 (Yippee-like 4) during murine erythroid differentiation. We found that Ypel4 is expressed specifically in terminally differentiating erythroid cells in both fetal liver and bone marrow, and is significantly upregulated as terminal erythropoiesis proceeds. Knockdown of Ypel4 expression by shRNA revealed that it is required for at least two critical events of erythropoiesis, induction of Ter119 antigen (a surface marker of terminal differentiation) and subsequent enucleation. Our present findings suggest that Ypel4 plays a necessary role in the terminal differentiation of red blood cells.

←— State aim of study.

Highlight the data that provide the best evidence for your conclusion. Be as quantitative as possible.

←— Incorporate methods with data description.

End with implication of research; the conclusion should resonate with the aim of the study.

Revise for Coherence and Concision

Because most readers of your paper will only view the title and abstract, it is extremely important that the paragraph is written well. Review Chapter 6 for advice on style. In the two versions of an abstract below, we highlight elements that improve coherence and concision.

Needs Improvement

Erythropoiesis, or red blood cell development, is an essential process for any animal that has a circulatory system. Terminal erythropoiesis is a complex yet very well

←— The introduction is lengthy and oddly does not reveal the actual goal of the project.

coordinated process that involves many cellular changes, including decrease in cell size, change in cell shape, increase in chromatin condensation and eventual enucleation, increase in hemoglobin synthesis, and changes in cell surface receptor expression. Mechanistically, the signals and processes that enable erythroblasts to undergo this transformation are just being understood. We previously identified Ypel4 (Yippee-like 4) as a gene of interest based on its high induction during terminal differentiation of murine erythroid cells. The Ypel4 gene is a member of the highly conserved Ypel gene family that encodes putative zinc-finger-containing proteins present in diverse eukaryotic organisms. Several of the Ypel family members are involved in processes central to the maintenance of life on a cellular level, including cell cycle regulation, growth inhibition and senescence, and cytokinesis. Here, we demonstrate that in mice, Ypel4 is expressed in terminally differentiating erythroid cells in both fetal liver and bone marrow; significantly lower levels of the RNA were found in granulocyte, monocyte, B cell, and T cell populations. In addition, Ypel4 is expressed nearly 1000-fold greater in late-stage vs. early-stage erythroid-specific cells, suggesting that Ypel4 expression increases as terminal erythropoiesis proceeds. Finally, knockdown of Ypel4 by shRNA revealed that it is required for at least two critical events of erythropoiesis, induction of Ter119 antigen (a surface marker of terminal differentiation) and subsequent enucleation. Our present findings suggest that Ypel4 is essential for hematopoietic development and further indicate the Ypel gene family as important regulators of eukaryotic life.

Information is so general that few readers will find anything new.

The broad topic of the paper is unclear as well: is the paper on erythropoiesis or the *Ypel* family?

The data are described with lots of detail, but details such as these do not seem necessary.

The last sentence does not resonate with the first sentence, but that could be a side effect of the unclear goal.

Better

The Ypel4 (Yippee-like 4) gene is a member of the highly conserved Ypel gene family that encodes putative zinc-finger-containing proteins present in diverse eukaryotic organisms. Although several of the Ypel family members are implicated in cell cycle regulation, the function of Ypel4 is still largely unknown. Here, we demonstrate that Ypel4 plays a necessary role in the terminal differentiation of red blood cells. In mice, Ypel4 is expressed specifically in terminally differentiating erythroid cells in both fetal liver and bone marrow, and is significantly upregulated as terminal erythropoiesis proceeds. Knockdown of Ypel4 expression by shRNA revealed that Ypel4 is required for at least two critical events of erythropoiesis, induction of Ter119 antigen (a surface marker of terminal differentiation) and subsequent enucleation. Our present findings suggest that Ypel4 is essential for hematopoietic development and further implicate the Ypel gene family as important regulators of eukaryotic life.

Relevant information about Ypel4 more effectively sets up question addressed by project.

Highlighting lack of information justifies the project.

Statement implies goal, but goal is clear: to define function of Ypel4.

Results are shorter, but still adequately describe evidence.

End resonates better with beginning of abstract.

The revised abstract more clearly states the goal of the project because the introduction focuses squarely on the gene *Ypel4* and because the justification is explicitly stated. Because the goal is clearer, the last sentence resonates better with the first sentence. Finally, the results are condensed considerably but still provide adequate evidence for the overall conclusion or argument.

Edit

Concision may be paramount in an abstract, but accessibility and clarity are, too. Here are a couple of things to seek as you edit your abstract.

Use the Correct Verb Tense

Use past tense for things you did or observed, but present tense for biological facts such as the implications of your research.

> Knockdown of Ypel4 expression by shRNA **revealed** that Ypel4 **is** required for at least two critical events of erythropoiesis.

Explain Abbreviations and Avoid Chemical Formulas or Jargon

Like the title, this advice depends upon your expected audience. For example, DNA does not need to be spelled out, but Ypel4 in our sample abstract does.

ABSTRACT CHECKLIST

Do

✓ Write an abstract that features key results but also includes the *aim* and *impact* of the work.

✓ Write concisely: every word counts.

✓ Use abbreviations and technical terms judiciously.

✓ Structure your abstract like a research article: Introduction, Methods, Results, and Discussion.

Don't

✓ Include too much background information or methods.

✓ End your abstract in a different way than your Discussion section: the end of each should resonate with but not replicate the other.

✓ Use improper verb tense: past tense for what you did/observed, present tense for biological principles.

✓ Include citations to illustrations or published literature.

Acknowledgments

Whether you are the only author or have coauthors, there are almost certainly other people who have given significant assistance to your project. The Acknowledgments at the end of the paper is the place where you can record how these people helped you. This is not the Academy Awards, where you recognize everyone you know: the Acknowledgements should include only those who have made substantial contributions.

Who should be included in this group? It differs from paper to paper, but typical categories include the following:

- People who gave you biological materials (e.g., cell lines, mutant stocks, antibodies, recombinant DNA constructs) that they have made themselves (i.e., not commercially available lines, stocks, or constructs). Alternatively, the donors can be listed with their gift when the material is described in Methods.
- People who allowed you to use an important instrument in their laboratory, provided help with statistical analyses, or gave you advice as you learned something new (e.g., a technique, instrument) for your project
- People who provided unpublished data that are crucial for your experiments
- People (including anonymous referees) who read drafts of your manuscript and gave you feedback. You could name specific individuals or thank an entire group.
- Sources of funding

Sample Acknowledgments

We thank Y. Hunt (University of Lisbon) for unpublished results and for the gift of the transgenic *Drosophila melanogaster* cell lines expressing the mouse p53 protein; B. Langer for allowing us to use her scanning electron microscope; D. Thehal for helpful discussions on protein purification; S. Tistical for bioinformatics advice; and members of the Y.N.H. laboratory for

Follow each donor with the name of the institution, unless the donor is from the authors' institution.

Specify gifts and services.

critical reading of the manuscript. We also thank two anonymous reviewers for their constructive comments on the manuscript. This work was supported in part by the National Institutes of Health under award number RO1GM123456 (to Y.N.H.). T.S.A. is the recipient of a Williams Young Investigator Award.

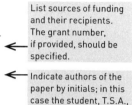

List sources of funding and their recipients. The grant number, if provided, should be specified.

Indicate authors of the paper by initials; in this case the student, T.S.A., and the professor, Y.N.H.

ACKNOWLEDGMENTS CHECKLIST

Do

✓ Acknowledge mentors and other people who contributed to your project in a substantial manner by providing reagents, instruments, and data.

✓ Specify the agency or group that awarded funding.

✓ List the institutions of people who do not work at your own institution.

Don't

✓ List everyone you know.

✓ Write more than one paragraph.

Supplementary Information

Supplementary Information (SI) or Supplementary Online Material (SOM) contains information that is relevant to a published paper but not necessary to the message of the paper. SI is peer-reviewed for quality and relevance to its published paper but is not included in the printed version. Instead, SI is linked to the electronic version, although it is cited in the printed paper (e.g., "see Figure S2") so that the reader who wants the extra information is able to find it.

What goes into SI? That varies widely from paper to paper. Typical examples include:

- Movies of results that printed journals can only show as static images
- More details of a procedure that is briefly described in Methods but has not been previously described in the literature
- Extra (i.e., less important) control experiments
- Very large data sets, so that they are easily available to the scientific community. The data in the SI spreadsheets can be used to evaluate the conclusions of the paper but, more importantly, can be mined by other scientists to answer other questions.

SUPPLEMENTAL INFORMATION CHECKLIST

Do

✓ Include material that may appeal to a smaller group of researchers, such as additional details about methods or data.

✓ Cite supplementary information in the main research article.

Don't

✓ Include material that is crucial for the research article: the research article must not rely on material that can only be obtained through the internet.

Before You Submit the Paper

Is the whole greater than the sum of its parts? You have carefully polished each of the sections of your paper and placed them in the proper order. Now is the time to take a break and then read the paper from beginning to end. A good way to do this is to read as though you were reviewing it for a journal; see the **Appendix: Reviewing like a biologist.** We assume you will pass this test and can submit your paper with confidence.

STRATEGIES FOR THE LABORATORY REPORT

3

Laboratory reports, like research articles, present the story of a research project in the standard format for writing about biological research: Introduction, Methods, Results, and Discussion. Compared to research articles, however, laboratory reports present a narrower scope of work. A typical research article describes several experimental approaches that together give a clear answer to the question under study, and frequently involves the collaboration of several scientists over a significant period of time. In contrast, a laboratory report often discusses a single experiment conducted by one or two scientists in a shorter time, frequently under time constraints that do not allow repetition of experiments.

Many of the differences between these two genres are due to differences in their intended audiences. Research articles are written primarily for scientists who are generally knowledgeable about the field, but are not necessarily familiar with the specific project described in the article. In comparison, laboratory reports are typically written for people with special knowledge of work related to the report. These might be members of the author's research group, members of another group collaborating on the project, or the professor and students of the teaching laboratory in which the experiment was done.

Although both laboratory reports and research articles are derived from information in laboratory notebooks, laboratory reports are the more closely related to laboratory notebooks because they are written directly from the notebook soon after the experiments are finished. In contrast, research articles may be compiled from several laboratory reports of experiments conducted at different times.

Laboratory Notebooks

Laboratory notebooks provide the experimental results for research articles and laboratory reports because notebooks are real-time records of each experiment, written while it is being carried out. Each entry is written primarily to aid the memory of the person doing the experiment but must be clear enough for others in the laboratory to be able to interpret the information. Because these records can provide evidence in decisions about intellectual property, laboratories frequently have specific regulations about the type of notebook used and whether entries are signed each day. Even if your laboratory does not make suggestions about the physical form of your notebook, don't yield to the temptation to take notes on paper towels with the intention of writing a neat notebook later. You can be sure information will be lost or forgotten.

The notebook contains the date of each experiment, the question that is being asked, and a detailed account of the materials and methods used. All of this information should be entered before starting the experiment, giving you a carefully thought-out plan and at least the beginning of a neat notebook.

continued

continued

The notebook also contains the raw data obtained from the experiment as well as unexpected observations and suggestions for other experiments. Therefore, each page should have space to make notes during the experiment about things that did not go as planned, accidents, or mistakes. Taking careful notes of unexpected events is extremely important; the unexpected often turns out to be more important than the experiment itself. One well-known example is Dulbecco's discovery that UV-inactivated phage could be reactivated by visible light (Dulbecco 1949). The discovery was made after irradiated phage had been accidentally left on a table under a fluorescent lamp. Apparently this photoreactivation was not noticed by other scientists because they never exposed the irradiated phage to visible light.

How to Use this Chapter

We have organized this chapter to explain how a successful student or a professional biologist would go about putting together a laboratory report. Because laboratory reports are primarily intended for a known audience—a research group or laboratory class—they frequently have requirements about formatting, nomenclature, or methods of analysis that are specific to that research group or class. To find out about these you should consult your group leader or instructor.

In this chapter we will describe a typical report as a basis for any modifications you are expected to make. All examples are from a student lab report on American chestnut trees that are resistant to chestnut blight. This disease has severely reduced the abundant population that existed in eastern North America before 1900. For this project, students developed hypotheses about environmental conditions that affect the health of surviving chestnut trees, and tested their hypotheses by analyzing American chestnut trees in southeastern Massachusetts.

Preparing Your Report

Most students faced with writing a laboratory report start by following the structure of the laboratory notebook: expanding on the experimental question for the Introduction, the lab protocol for the Materials and Methods section (hereafter, Methods), their observations for the Results section, and their interpretations for the Discussion. This process got them to their final results, but it is not the most effective way of helping the reader to understand the conclusions and their relationship to various pieces of data collected along the way.

Most experienced biologists instead begin to write by organizing their final results and then build their report draft around these results so that the reader is continually aware of how the parts of the story are related. We suggest that you likewise begin with Results: start by organizing the figures and tables (hereafter, collectively referred to as illustrations) that present the data. Arrange those illustrations in the order that best tells the final story (not necessarily the order in which the experiments were done). The Results, in turn, help organize the Methods section, which should parallel the experiments discussed. Only after drafting the Results and the Methods sections do most experienced lab report writers compose (or modify) the Introduction to give the reader the necessary background information and complete the Discussion to consider the implications of the findings. Below we discuss details of each section in the order described in which they are typically written. This approach is identical to that used for research articles, discussed in Chapter 2. If you still have questions about specific sections after reading the descriptions below, we suggest you consult the equivalent section in Chapter 2, which contains more extensive details.

Although the process of preparing a laboratory report is basically the same as that of preparing a research article, there is one notable difference. Research articles are usually limited by the word counts and space a journal will allow, requiring hard choices in which illustrations most efficiently tell the story. Laboratory reports are not as constrained for space as research articles; thus, you do not have to eliminate any illustrations that support your story. While for many students a lab report may be a stand-alone assignment, remember that for practicing biologists, the laboratory report represents the first complete, organized document of results from different parts of the project. Such reports are the records that experimenters actually use when interpreting data and designing the next set of experiments. Therefore, both the Methods and Results sections of a laboratory report should include all the important details.

Figures and Tables: Start with Your Data

Figures and tables are used to present data in an organized manner. There are three types of figures (graphs, images, and diagrams) and you should choose the type of figure that best conveys the message of your data. We include examples of some types. More details about figures and tables can be found in Chapter 2.

Graphs are useful tools for understanding the significance of numerical data. There are numerous ways to graph data (see Chapter 2) and you may want to try more than one to see which best displays your result. In general, bar graphs work well for comparing a small number of samples, and line graphs reveal trends with time or distance (Figure 3.1).

Images are photographs or diagrams of biological material to display the physical structure or to show the spatial relationship between the objects of study (Figure 3.2).

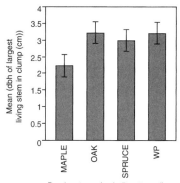

FIGURE 3.1 Relationship Between the Dominant Canopy Tree
Species and the Mean Diameter at Breast Height
(DBH) of the Largest Living Stem of the American
Chestnut. "WP" indicates white pine, and "maple,"
"oak," and "spruce" indicate any member in that
Genus. Each error bar indicates standard error of
the mean.

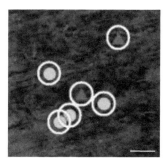

FIGURE 3.2 Section of a Map Illustrating the Distribution of
American Chestnut Trees on Water Tower Hill.
Triangles indicate diseased trees. Solid circles
indicate non-diseased trees. The size of the outer
circle represents the percentage of canopy cover
over each tree. Bar = 50 ft.

TABLE 3.1 Characteristics of American Chestnut Sprouts
and Their Environments

Site- EB or WT	N Rows	# Shoots in clump	DBH of largest living stem (cm)	DBH of largest dead stem (cm)	Distance of nearest tree (m)	# of total trees in 4m radius	leaf litter depth (cm)	% canopy cover
EB	18.00	4.89	3.59	5.71	1.71	6.83	6.28	68.06
WT	28.00	5.54	2.78	4.72	1.61	7.46	6.97	65.89
Each number represents a mean value. Elm Bank (EB) and Water Tower Hill (WT) are located less than 4 miles apart in southeastern Massachusetts.								

Summary tables are useful for simplifying findings. Often these data are grouped into categories (e.g., chlorotic vs. green, high vs. low) that are not appropriate for statistical analysis.

Tables help organize data from many samples, especially if each sample has several types of measurement. You probably recorded your data in a table (Table 3.1) long before you wrote your report. Tables are also useful in the Methods section to give characteristics of study sites or biological material used in the research (Table 3.2).

Legends for Figures and Tables

Each figure and table needs a number and an informative title that describes the illustration. Figures and tables are numbered separately and in the order in which they are first cited in the manuscript.

Most figures and tables also need a legend that enables the reader to interpret the illustration without reading the description in the text. Therefore, your legends should explain symbols and abbreviations used in the illustration, list samples, and mention the assay used to obtain the data. The techniques themselves do not need to be described in detail because this information would be more appropriate in the Methods section.

FIGURES AND TABLES CHECKLIST

Do

✓ Choose an illustration that most effectively conveys the point of the data.

✓ Write a legend that gives the information needed to interpret the illustration: describe how the data were generated, briefly list samples, and explain abbreviations and symbols.

✓ Number tables separately from all the other illustrations. Both tables and the other illustrations should be numbered in the order in which they are cited in the lab report.

Results: Connect Illustrations and Text

The Results section can be compared to a travelogue. Your figures and tables are the photographs you have taken on your trip; the text is the narrative you write to link the photographs into a story (not a random set of photos). The Results section should begin with a brief paragraph stating your aim and experimental strategy of your project. Following that, each experiment should be described in a separate paragraph, starting with a clear topic sentence that introduces the rationale for doing that experiment. If your Results section is short, such topic sentences may sufficiently guide the reader; you probably will not need the kinds of subheadings typically found in a research article. The paragraph then cites the relevant illustration(s), directs attention to the aspects of the data that relate to the question being asked, and ends with a brief conclusion that provides the context for the next experiment.

The text in the Results should describe the data clearly enough for the reader to understand it without referring to the illustration, just as the legend should allow the reader to interpret the illustration without looking at the text. However, what is required in the text differs from what is required in the legend. The text explains why the experiment was carried out and what conclusions can be drawn, while the legend highlights details of the experiment and allows readers to evaluate the quality of the data. In terms of a travelogue, the text has a more general description of the experiment in the illustration, analogous to what you would say about your hike up the mountain if you had no photos. Why did you choose that trail? A friend recommended it. What did you learn on the trail? Although there are many trees, the rocky ledges gave dramatic views of the valley and lake and there is evidence of the early mining activity. What did you decide about the trail? It is the most interesting trail we have seen on this trip. The legend, on the other hand, has the specific details that you would give when showing a photo. Where was the photo taken? About halfway up the trail. What is that wooden structure? The entry to an abandoned mine. Who are the people? Two hikers coming down the trail.

Sample Results

This report examines factors that contribute to blight resistance of American chestnut trees (*Castanea dentata*) in two study sites, Water Tower Hill and Elm Bank. In both sites chestnut trees have undergone the phenotypic change seen across the country and now reproduce, not by scattering nuts, but by sprouting from roots so that each tree becomes a clonal cluster. Considering the largest sprout in each cluster the focal stem, we mapped each Chestnut cluster using a handheld GPS unit and ArcMap software. The infection status was determined by

← Reiterates the aim and experimental design of the project.

ranking the type of cankers. For each cluster we then measured two aspects of the local physical environment as well as the dominant species of their neighboring trees.

The mean depth of leaf litter surrounding a chestnut is one aspect of the local environment that has been reported in previous studies (Tindall et al. 2004). We measured litter depth at 5 points on a four meter radius around each tree (Figure 1). Values for diseased trees were then compared to those for non-diseased trees. Although the number of diseased trees appeared to increase with mean litter depth, we found no significant statistical relationship between these two factors (Wilcoxon, $x^2 = .4203$, df = 1, p = .5168).

Explains why they performed the experiment and supplies a reference to a relevant published study.

Conveys the conclusion of each illustration and describes the supporting evidence.

Another aspect of the local environment studied was the percentage of canopy cover, which affects the amount of light available to the plant. We estimated this by visual estimation of a ten meter circle above the tree (Figures 1 and 2). Although it appears that the incidence of disease decreased as canopy cover increased, we found no significant statistical relationship between these two factors (Wilcoxon, $x^2 = .902$, df = 1, p = .3425).

The local environment is influenced in several confounding ways, such as canopy cover, by neighboring trees. The nearest dominant canopy tree species was recorded within a five meter radius around each chestnut cluster. We then determined the mean diameter at breast height (DBH) of the largest living stem of the chestnut, as an indicator of fitness (Figure 3). Although this fitness appeared to decline in an area dominated by maple trees, a Tukey test showed no significant statistical difference between the effect of any species on the fitness of the American chestnut.

Cites the relevant illustration.

RESULTS CHECKLIST

Do

✓ Create a logical narrative based on the illustrations.
✓ Cite relevant figures and tables.
✓ Describe your data adequately so that a reader does not need to look at an illustration.
✓ Include rationale and conclusion for each experiment.

Don't

✓ Report irrelevant or inappropriate results (i.e., results not related to the question being studied).

Methods: Document Your Process

The Methods section of a laboratory report, like that of a research article, serves two purposes: (1) it allows other scientists to replicate the experiments and (2) it provides specific experimental details, which affect how readers interpret data in the paper. However, the different audiences for these two genres dictate significant differences in the level of detail.

Most research articles do not provide enough detail for a reader to exactly reproduce many of the experiments because readers who want to replicate an experiment usually want to replicate it with their own biological material (e.g., to see whether what is true of the mouse is also true of the elephant). In contrast, laboratory reports, whether written for a laboratory class or a research group, are frequently intended for exact replication. For example, your study of the kinetics of a particular enzyme reaction may be compared with those of other students in the class who are doing the same experiment with mutant versions of the enzyme. In such cases it is important that everything that can affect the reaction be recorded in the laboratory report so that these factors can be kept constant by other experimenters. This does not mean, however, that you repeat every detail of your protocols or laboratory manual; see Chapter 2 for a comparison of the two. Consult your instructor or other laboratory reports to determine the level of detail expected in your report.

With the exception of the laboratory-specific requirements for level of detail, the Methods section of the laboratory report is written in the same way as the Methods section of a research article. In both cases the goal is to turn protocols (recipes) into a story that helps the reader to understand how the data were generated. Include information on the biological material and other reagents, how you treated them, and how you analyzed the data. Organize the procedures using the same order in which they appear in your Results. Just like the Results section, the use of subheadings is optional in the Methods section of a lab report because of its short length. Finally, use precise wording and write unambiguous sentences so there is no question of exactly what was done. Because these goals are approached in the same ways in both laboratory reports and research articles, we suggest you read the section **Methods: document your process** in Chapter 2.

Sample Methods

This Methods section includes several subheadings, which are not always necessary in a Methods section but can help a reader find information about a particular procedure. If you use subheadings, they would vary depending on what you are studying and the nature of the experiment.

Study site. The distribution of regenerated American chestnut sprouts and the status of their infection from Chestnut blight (Table 3.2) were investigated in two forested areas within four miles of each other on October 7, 2013. Water Tower Hill (WTH) is

> Start your Methods section by describing the most basic component. Here, it is the trees and their environment.

a drumlin in Wellesley, Massachusetts, and Elm Bank Reservation (EB) lies along the banks of the Charles River in Natick, Massachusetts, and is relatively flat and lower in elevation.

Tables can efficiently display characteristics of material studied.

TABLE 3.2 **Summary of Chestnut Stems Affected by Chestnut Blight at Two Sites in Southeastern Massachusetts**

Site	Number of Diseased Trees	Number of Not Diseased Trees	% Diseased
Elm Bank	7	11	39
Water Tower Hill	16	9	64

Mapping infected chestnut stems. The location of clusters of regenerated chestnut stems was recorded using Garmin E-Trex GPS receivers according to the Universal Transverse Mercator (UTM) co-ordinate system. Within each cluster, the largest stem was classified as either living or dead and was designated as the focal point from which a circular plot with a radius of 5 meters was created. Fitness of the stems was evaluated by measuring the circumference of the diameter at breast height (DBH) of focal stems, counting the number of stems within the clump, and determining the type of canker. To assess the severity of infection, the cankers on the largest stems were ranked as either 1) not visible, 2) sunken or 3) swollen.

Provide enough detail about how data are gathered and analyzed, but do not regurgitate your protocols.

Determining leaf litter depth and percentage canopy cover. The resources within the circular plots were characterized at both sites by evaluating the depth of leaf litter and availability of light. Within each plot the leaf litter depth in the O horizon layer of the soil profile was recorded and averaged from 5 randomly selected places. The forest canopy tree nearest to the focal stem was identified and the distance to the focal stem was recorded.

The percentage of canopy cover over the cluster was visually estimated.

Data analysis. Differences between the sites were compared by analyzing the number of stems per cluster, the type of canker, DBH and the leaf litter depth. Variables such as canopy cover, litter depth, and the species of neighboring trees were also analyzed using JMP statistical software (version 10). The relationship between spatial location and variables including the type of canker or number of stems per cluster were investigated and visualized with ArcMap (ArcGIS: ESRI).

METHODS CHECKLIST

Do

✓ Describe your components (biological material, other reagents and materials), conditions, and data processing.

✓ Provide enough detail so that a biologist can repeat your experiments.

✓ Organize subsections and paragraphs in a logical manner: biological material first, and after that the order of methods should parallel the order of the Results.

✓ Write precisely.

Don't

✓ Regurgitate your protocols.

✓ Write in strictly chronological order.

Introduction: Start Broadly and then Narrow to Your Aim

Like the Introduction of a research article, the Introduction of a laboratory report includes the aim, justification, and context of the project. The major difference for the lab report is the context: the Introduction of a lab report is not as extensive as the Introduction for the research article. Why? First, the goal of the lab report project is narrower in scope—perhaps a project in calibrating instruments or in repeating a well-known experiment. Second, the audience—typically, your professor or teaching assistant—is already familiar with the specific work. Finally, there may be insufficient time for an extensive literature search. It's not uncommon to have only a day in the lab for a lab report project. In such cases, your professor may provide one or two references to help you write the Introduction. Because of the narrower context, the Introduction of a laboratory report can be written in one or two short paragraphs. Nevertheless, the Introduction should begin with a broad statement of the general context before narrowing to lead the reader to the justification/motivation and aim of your study. For more information on developing the context from your justification and aim, see Chapter 2.

Sample Introduction

In 1904, an invasive fungal pathogen (*Cryphonectria parasitica* (Murrill) Barr) spread throughout North America, causing detrimental effects on the American chestnut (*Castanea dentata*) population. Within 3 decades, the blight infected and killed over 3 billion chestnut trees, changing the ecological functioning of many forests in which chestnuts had been the dominant species. The American chestnut is now considered functionally extinct, but scientists have studied methods to reintroduce blight-resistant American chestnut back into its natural habitat. In order to successfully reintroduce the chestnut, it is important to understand the environments in which American chestnuts continue to

← Start your Introduction with a broad statement to frame your context. Here, the student gives a historical perspective.

← Provide the justification for your study.

survive despite being exposed to the blight for nearly a century.

American chestnuts that survive are significantly smaller than chestnuts before the blight, and now tend to be a marginal component of the forest understory. On a larger scale, Burke (2011) found that populations of surviving chestnut trees are restricted to a portion of their original niche before the introduction of the blight. These niches where the American chestnut now persists can help us understand the factors that influence the fitness of the species.

In order to better understand the effects of neighboring species and environmental conditions on the health of the American chestnut, we studied a population of surviving trees in the northeastern United States. We explored two hypotheses: 1. Environmental conditions predetermined by the location, particularly leaf litter depth and canopy cover, have an effect on the health of the American chestnut. 2. The dominating species neighboring the American chestnut have a major effect on the health of the tree.

← The Introduction to a laboratory report typically can be written in one or two paragraphs, but this student uses a third paragraph to highlight the aim of her study.

INTRODUCTION CHECKLIST

Do

✓ Explicitly state the aim and justification of your project.
✓ Develop context that is relevant to your goal and appropriate for your audience.
✓ Create a funnel-like overall structure: start broadly, and end with your goal.

Don't

✓ Include unnecessary background or repetition in your context.

Discussion: Interpret Your Data

The purpose of the Discussion section is to interpret your results. What have you found and how does it relate to the work of others? How would you have designed the experiment if you had known what you know now? If your findings are equivocal, what are the limitations of your approach? How could you improve your approach? If your findings support a completely unexpected answer, why did this happen? Of course, this result could be due to human error, but much interesting science has come from pursuing an apparently failed experiment, so carefully consider other reasons for unanticipated data.

As with the other sections, the limited scope and specialized audience of the laboratory report affect the focus of the Discussion section. The Discussion section of a laboratory report tends to elaborate on details of the experiment itself (e.g., why things went wrong) rather than the broader biological implications discussed in research articles. However, connecting your work with those of others (e.g., in the laboratory course, from the published literature) can be done and would enrich the content of the Discussion section. In some laboratory courses, students are not expected to search the literature but may be given a reference or two to help them develop their Discussion section.

Sample Discussion

The persistence of regenerated American chestnut sprouts in our New England forests allows us to continue to study the environmental conditions that might lead to increased fitness of these trees. We evaluated three aspects of every plot around each cluster: depth of leaf litter, percent canopy cover, and species of nearest canopy trees. We found that none of these factors significantly affected the fitness of the American chestnut in our study.

Opening paragraph reminds reader of justification and aim of project. The paragraph ends with the overall argument of the study.

Our study, however, gives a slight indication ← Body paragraph interprets results within the context of previous findings.

that maple trees adversely affect the health of American chestnuts. American chestnuts in maple-dominated environments exhibited a smaller DBH than those that grew in environments dominated by other tree species such as Oak and White Pine (Figure 4). The negative influence could be due to the fact that both the sugar maple (Schwadron, 1995) and American chestnut (Wang et al., 2006) are shade-tolerant species, leading to increased competition for resources and shade. Our observation could be further explored by analyzing the interaction of the effect of the dominant tree with other environmental factors.

In summary, the blight-resistance of ← Closing paragraph reiterates overall conclusion and impact of study.

American chestnut trees in Natick and Wellesley, MA, was not found to be significantly associated with depth of leaf litter, percent of canopy cover, or species of neighboring trees. Our work, however, provides a dataset of American chestnut trees that can be used to consider interactions of multiple environmental factors. Such studies will help identify the conditions necessary for restoring the American chestnut.

DISCUSSION CHECKLIST

Do

✓ Begin by reminding the reader of what you did and why, and end by reiterating the project's overall conclusion and impact.

✓ Explain your contribution to the field by comparing your work with those of others.

✓ Include limitations of your methods and interpretations.

Don't

✓ Regurgitate the Results section: highlight only the data that support your claims, and avoid redescribing the data.

✓ Write arguments that are wordy, unclear, meandering, and so forth.

Title

The title of a laboratory report should be a concise and accurate reflection of the content of the report and should address the original questions. Because these reports are unpublished and written for a known audience, titles are free of two major constraints faced by research articles (see Chapter 2): they are not limited in length and they have more freedom to use laboratory-specific terms that might not be understood by the broader audience of a published article. For example, the title below is not only concise but also specific about the organism and the agents being studied.

Effects of local environment and neighbor- ← Variables
ing species on surviving populations of
the American chestnut (*Castanea dentata*) ← Organism

TITLE CHECKLIST

Do

✓ Express the content of the report as a concise, informative phrase.

✓ Include specific information such as the organism being studied or the variables analyzed.

Don't

✓ Be vague about the content of your work.

✓ Be wordy.

References

Although laboratory reports typically do not have long lists of references, every paper that is cited in the report must be listed in the References section at the end of the report. Chapter 7 discusses several options for formatting Reference sections and the citations in the text. Consult your instructor or laboratory director for the preferred format for your laboratory report.

Appendix

Laboratory reports do not usually suffer the space constraints of published papers; however, in some cases these reports do need to include large sections of information that are of interest to only a fraction of readers. These large sections include some of the things that are found in the Supplemental Information (SI) of research articles. For instance, a list of the GPS locations of every tree you studied with the tree's infection status could be used by a scientist who wanted to study another aspect of that population. Such lists are often put in an appendix to make the body of the paper easier to read.

APPENDIX CHECKLIST

Do

✓ Include material that may appeal to a smaller group of researchers, such as additional details about methods or data.
✓ Cite appendix information in the main research article.

Don't

✓ Include material that is crucial for the laboratory report: readers assume that the appendix is only for those with a special interest.

STRATEGIES FOR
LITERATURE REVIEWS

4

A **literature review** surveys the primary literature to provide
an up-to-date understanding of a specific subfield, topic, or
question. Literature reviews serve many different purposes
and come in different formats (Table 4.1) such as the Introduction of a research article discussed in Chapter 2. Readers typically use a literature review to become familiar with a new
field—its history, terminology, present consensus, and current controversies that warrant further investigation. Funding
agencies require literature reviews in grant proposals to ensure
that the proposed project is sufficiently significant and innova-
tive (i.e., not reinventing the wheel). Finally, writers of literature reviews benefit from the opportunity to step back from
the minutiae of their experiments and obtain a broader view of
the field: they may be able to make connections that weren't
obvious before. Instructors may assign literature reviews for

TABLE 4.1 Common Types of Literature Reviews

Type	Examples	Purpose	Length[a]
Stand-alone full-length review	*Annual Reviews*, *Nature Reviews*, the *Current Opinion* family of reviews	Provides compre-hensive introduc-tion to subfields	2,500–10,000 words

Stand-alone mini-review	Commentaries in *Proceedings of the National Academy of Sciences of the United States of America*; News & Views in *Nature*; Perspectives in *Science*; Previews in *Cell*	Advertises specific articles in the journal, providing a more general background to increase accessibility to people outside of the field	750–1,000 words
Part of a research article or grant proposal	Introduction section of research article; Significance section of grant proposal	Gives the relevant background information to explain and justify a specific project	500–750 words

a For all types of reviews, the lengths vary greatly depending on where they are published and the topic being reviewed. This table gives only rough estimates to allow comparisons of the three types of literature reviews. Similarly, literature reviews assigned in courses will differ in length depending upon the model chosen by the instructor.

learning purposes: to make students aware of available research and become more critical readers of it.

Focus Your Topic

As you start a literature review, keep in mind the scope of your search, and modify the breadth accordingly. Let's say that you have been asked to write a 2,500-word literature review on some aspect of cancer. As you can see in Figure 4.1, there are many aspects (e.g., mechanism of the disease, drug treatment) to consider. So, you choose to focus on cancer drugs, but again, there are many aspects your literature review can cover, such as trials for a specific drug, and its costs. Even though you decide to home in on a very specific topic, such as the mechanism of the cancer drug imatinib, consider citing a few articles that talk more broadly about the disease (e.g., the societal impact of cancer) to help establish the importance of your topic.

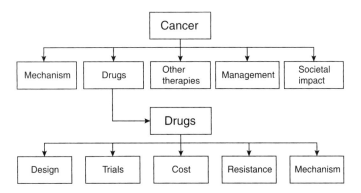

FIGURE 4.1   Levels at which Searches Can be Conducted. Such
a flowchart can also help you structure the body
of your literature review.

Locate Sources by Using the Right Search Engines

Literature reviews synthesize original research, so you should
build your bibliography from the primary literature. The primary literature is vast (scientists are rather prolific writers),
but fortunately a number of search engines can help you quickly
find the most relevant articles for your review.

Specialty search engines exist for the many disciplines
within biology. We list a number of them in Table 4.2; consult
the librarian of your biology library for available search engines. A particularly useful tool is the Cited Reference Search
of ISI's Web of Knowledge site. Although the Web of Knowledge is a broader database (covering social sciences and the
arts and humanities in addition to the sciences), the Cited Reference Search tool gives you a list of articles that reference a
particular citation.

Depending upon the number of journals to which your
institute subscribes, it should be relatively easy to access the
full-length articles at no cost to you either through your institute's subscription or through the database. If, however, you

TABLE 4.2 Databases Commonly Used in Biology

Database	Biological Subfield
ASFA	Aquatic science
AGRICOLA	Agriculture
BIOSYS	Biological and biomedical sciences
CAB Abstracts	Applied life sciences
Ecology Abstracts	Ecology and environmental science
POPLINE	Reproductive health and development
PubMed	Biomedical science
Web of Knowledge	All areas of science, social science, arts and humanities
Zoological Record	Zoological and animal science

cannot obtain the articles, do not despair! Your institution's library can easily obtain the article for you through its interlibrary borrowing system. The trick is that it takes time to fulfill requests (perhaps a week or more), giving more incentive to start your literature search early.

What To Do About Google And Wikipedia?

Two popular search engines, Wikipedia and Google Scholar, are fine places to start when you are completely new to a topic but are not appropriate for finding the bulk of your literature. Wikipedia can familiarize you with key terms and ideas of a field, but the reliability of the information is questionable because it is not peer-reviewed. Google Scholar can help you quickly identify the most influential papers because search results are ranked by the number of citations, but Google Scholar does not provide as much access to the peer-reviewed journals as specialty databases do.

Search for Articles Strategically

The biggest problem we encounter in helping students write literature reviews is that students cannot find enough material on their selected topic. Most of the time, it is because the student lacks a clear research strategy. Here again, the biology librarian at your institute can help you identify useful terms, journals, or articles to get you started. Here is a sequence of steps you can follow on your own for an effective search.

1. **Pose a clear question** to help you focus your research. For example, what mutations in the BCR-ABL gene lead to resistance to the cancer drug imatinib? Do not, however, simply put this question into the search engine box! Doing so may narrow your search too much, and indeed, may give you no results at all.

2. **Identify the unique concepts** in the question. Some key concepts in our sample question are point mutation, imatinib, and BCR-ABL. Breaking down the question into key concepts helps you identify the terms that are most likely to give you the type of material you seek. You should not, however, limit your search to these terms.

3. **Make a list of terms** that are related to your key concepts to give you flexibility in your search. For example, some related terms for our question are Gleevec (a clinical name for imatinib), CML (for chronic myeloid leukemia, the disease caused by a mutant version of BCR-ABL), cancer, and drug resistance. Skimming even just one article on your topic will reveal key terms that you can use in later searches. An initial search can also provide ideas for better search terms. To use a nonscientific example, let's say you were conducting a web search for advice on creating scientific posters. Searching for the phrase

"scientific poster" will likely return calls for conference proposals, which only specify the size of the poster. Adding the term "effective" or "design" provides more relevant information.

4. **Manipulate the key search words and concepts** in ways that will give you even more flexibility and, we hope, more relevant results. You can use the symbols for wildcard (e.g., the term ?NA will search for DNA and RNA) or truncation (e.g., enzym* will search for enzyme, enzymology, and so forth). You can also combine terms via quotation marks (e.g., "chronic myeloid leukemia") or Boolean operators such as AND, OR, NOT. Consider, however, reserving Boolean operators until later in your search, and avoid searching for a long string of terms. It would be better to use several strings that combine different sets of key terms together. One search is hardly ever sufficient. Indeed, even seasoned researchers will conduct multiple rounds of searches, using various permutations of and variations on the key search terms.

5. **Mine the bibliographies of a few relevant articles** (e.g., a review article given to you by your instructor or supervisor). Those bibliographies are the result of scientists performing literature searches for you. Obtain the original articles to make sure you interpret the information in the same way as the authors who cited the articles.

6. **If you know the citation of the article, search for it directly.** A useful shortcut to find articles in PubMed is to type in the last name of the first author and the first page. For example, the phrase "Meselson 671" will return the article that describes the classic Meselson–Stahl experiment, which demonstrated the semiconservative nature of DNA replication (Meselson 1958).

How to Read Research Articles

How do professional biologists read articles?

To keep up with progress in their field, biologists constantly search databases for relevant articles and are then faced with deciding which of the many papers are the most important to read. Papers that are obviously relevant to the biologist's interest are read from beginning to end. But often the Abstract does not have enough information to indicate whether the paper contains information the reader is seeking. To speed the process of identifying appropriate papers, biologists depend on the organization of research articles (see Chapter 2). Because material is placed in predictable sections, readers jump to different parts of a paper as they decide whether the material is applicable to their work.

In deciding whether to read a paper, biologists frequently start reading the Discussion section, which gives the most information about the author's conclusions. If the conclusions are of interest, they then move to the Introduction before finally deciding whether to continue to Results, checking with Methods when necessary to clarify questions raised by the Results. Occasionally, a scientist very familiar with the field may simply analyze the illustrations in the Results section or check some technique in Methods.

How can you get better at reading research articles?

For inexperienced biologists, the goal is not to decide which research article to read, but instead to understand the unfamiliar science of the chosen paper. This process involves multiple reads to completely decode the language. If you are completely unfamiliar with the topic, start by reading a brief review on the topic. Despite its faults, Wikipedia is useful here because of its concision and broad coverage. The next step is to start reading the paper straight through, from beginning to end, several times. The first time through you should not stop and worry about the meaning of every unfamiliar word, concept, and experimental method—only those that prevent you from understanding the essence of the argument. Keep the other terms in mind, however, as you read other parts of the paper because you will find clues to help you decipher their meaning. Do not omit the Methods section, because learning how the experiments were carried out helps put the results in context. If the paper has Supplementary Information online, skim this also because sometimes methods or experimental material are explained in greater detail.

The second or third time you read the paper, you should find that some of your unknowns are beginning to make sense, especially if you do not wait too long between reads. When you reach this point, you should look up any words, methods, or concepts you still do not understand. Google and Wikipedia can be very effective for this task as long as you realize that many of Google's answers, especially the top answers, are probably not relevant to your question. When you feel reasonably comfortable with the paper you can begin to critique it.

Keep Track of the Literature

As you gather your sources, consider using citation management systems such as Mendeley and Zotero (see more information in Chapter 7). Such software can help you organize and format your bibliography and insert your citations as you write; some allow you to export citations directly from your database. Check your institution's library for the systems it recommends and supports.

Bibliographic software also allows you to take notes on articles. We recommend **annotating your articles** as you read to help keep track of what each article said. You can find examples of annotations in the secondary literature such as in the *Current Opinion* and *Nature Reviews* series of journals.

To annotate an article, write one or two sentences that summarize the article (e.g., the purpose, key result, impact) and state the relevance of the article for your literature review. Annotating articles not only helps you remember what you read, but may also help you to avoid plagiarizing them because you are describing the articles in your own words.

Kim BY, Jiang W, Oreopoulos J, Yip CM, Rutka JT, Chan WC. Biodegradable quantum dot nanocomposites enable live cell labeling and imaging of cytoplasmic targets. Nano Lett. **8**, 3887 (2008).

◄— Entry includes all the author names and full title as well as the journal name and volume, first page, and year.

This paper discusses the use of biodegradable polymeric nanospheres containing antibody-coated QDs which are internalized and degraded in the cell, releasing the QD contents which can then associate with molecular targets determined by the antibody coat. Noninvasive methods of delivering QDs

◄— The first sentence of the annotation summarizes the article.

to cells are needed in order for QDs to be useful for *in vivo* studies, where major disturbances caused by QD internalization could alter the function of the process under investigation. ← The second sentence explains the relevance to the review.

Another strategy to help you keep track of what you have read is to **make an evidence table** that identifies the key elements of the paper:

- Introduction: state goal and justification of the project in one or two sentences
- Method: describe key method(s)
- Results: describe key results in two or three sentences
- Discussion: state overall implication of research in one sentence

Table 4.3 shows an excerpt of an evidence table that summarizes the methods and results of three different articles. All articles describe findings about the mechanism behind a mutated BCR-ABL gene, but the scientists in the articles used different types of techniques for the experiments.

TABLE 4.3 **Sample Evidence Table**

Source	Methods	Results
Gorre et al., 2001	Biochemical analysis of clinical material	Resistance due to T315I in BCR-ABL or gene amplification
Roumiantsev et al., 2002	Biochemical analysis of clinical material	Resistance due to BCR-ABL mutations at Y253
Azam et al., 2003	*In vitro* screen of mutagenized BCR-ABL	New mutations reveal novel allosteric mechanism

Based on this chart, one can make the following statement that synthesizes the information:

> Imatinib-resistant mutations have been found in cells from cancer patients and after *in vitro* mutagenesis (Azam et al. 2003; Gorre et al. 2001; Roumiantsev et al. 2002).

This exercise not only helps you identify the salient feature of the articles, but also helps you see relationships between articles. Constructing an evidence table will also help you to paraphrase information because you describe the elements of the paper in your own words (see Chapter 7 for more advice on paraphrasing).

Know What You are Reading

A colleague of ours has a wonderful story to show why you should read the original article. The class project of her graduate biochemistry lab was to study enzyme X. A recent review cited many papers stating that this enzyme was found from bacteria to humans and listed Tetrahymena as an organism in which enzyme X had been studied but said its regulation was not known. Our friend and her lab partner were assigned to work on Tetrahymena. They obtained a culture and learned to make protein preparations. All went well until they tried to get enzyme activity from their *in vitro* preparations. Protocols for other organisms worked well for classmates, but none of these worked for Tetrahymena. The Tetrahymena paper cited in the review was old, written in German, and not in the Biology Library, so they had not read it. Finally, there was no recourse. They went to the Chemistry

Library, climbed a ladder to a top shelf, blew the dust off the book, and sat down with their German dictionary. Although the title of the paper translated to something like "The *X* enzyme in Tetrahymena," the paper actually described unsuccessful attempts to find this enzyme in Tetrahymena. The moral of the story? Be sure you know what the author means by the title (even if the person writing the review did not).

Know the Difference Between Summarizing and Critiquing the Literature

One of the challenges of writing a literature review in an unfamiliar field is that you have to read quite a lot in order to gain enough competency to identify the latest trends. Does that mean you have to read every single page you collected? No! Here's a little secret: scientists rarely read research articles from beginning to end; they only glean information from parts of the article. The Introduction and Discussion are the most relevant sections for writing a literature review because they discuss research articles in relationship to each other and the implications of the work—which is exactly what you want to do in your own literature review. These sections also happen to be the easiest sections of a research paper to read because they lack the technical details present in the Methods and Results sections.

Although reading the Introduction and Discussion sections will be sufficient for many of the articles you cite in your review, you will also need to evaluate articles that are seminal, play a key role in a controversy in the field, or are closely related to your own experiments. For these papers, you should concentrate your efforts on the Methods and Results sections—but you may not need to read the whole sections. Consider adding a "Comments" column to your evidence table for your notes as you think about these points:

1. **Identify the key experiments** (i.e., experiments that are crucial to the main argument of the paper). Often, the key experiments are mentioned in the Abstract of the article, so you just need to find the corresponding illustrations. Keep in mind, too, that a figure often has more than one graphic, so a key experiment may actually be part of a figure. Indeed, of the five to seven multi-paneled figures of a research paper, sometimes as few as three experiments build the argument. Other experiments are frequently controls for the argument being made in the paper; however, depending on the reason you are reviewing the literature, these controls may be more relevant to you than the other results.

2. **Evaluate the methods of the key experiments.** Are the techniques standard in the field? What are the limitations of the techniques? Resources to help you answer these questions include the internet, essays in the *Current Protocols* series, and advanced undergraduate biology textbooks. You can also evaluate the paper as a whole based on the number of techniques used: researchers strengthen their arguments when they use more than one type of technique (e.g., genetic and biochemical).

3. **Consider the quality of the data.** Were the proper controls used? Are the data statistically significant? While answering the latter question, take into account the organism used. Experiments with mammals (mice, humans) tend to have smaller sample sizes than experiments that use fruit flies or worms because mammals are more expensive to maintain.

4. **Evaluate the logic of the researchers.** Does the description of the data match the analysis? What other interpretations can be made based on the data? Even though papers have been reviewed by experts, it is quite possible that the data do not look convincing to you. The final question you want to ask is, how reasonable are the assumptions of the researchers? Some assumptions may be addressed in control experiments elsewhere in the paper; others may not be addressed, but perhaps should be.

The Literature Review is Not a Book Report

One of the biggest challenges of writing a literature review is avoiding writing a series of summaries. The problem with this is that the overall structure is determined by the individual articles. What you want instead is to organize the literature review by the relationships and patterns among your articles. As shown in Figure 4.2, a good literature review results from an interaction of three processes: summarizing, critiquing, and synthesis.

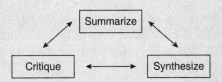

FIGURE 4.2 Processes Involved in Writing a Literature Review.

To illustrate the difference between critiquing and summarizing, let us revisit an experiment described at the end of the Results section of Chapter 2. This article

continued

continued

describes a study of the role of IcmF in the pathogenicity of the *Helicobacter bilis* bacterium. Compare these two statements that describe the result of the growth assay experiment:

Summary

In liquid media, *H. bilis* that lacked IcmF grew more slowly than wild-type *H. bilis*, suggesting that IcmF is important for bacterial growth.

← Describe information objectively.

Critique

Although IcmF was shown to be important for *H. bilis* growth *in vitro*, *in vivo* studies in mice are necessary to confirm the finding.

← Highlight a limitation or strength of the finding.

The first sentence simply describes the experiment, while the second sentence critiques it by highlighting a limitation of the procedure. The critique does not always have to be negative, however: you could also point out the strengths of an experiment.

Identifying trends necessary for synthesis can be helped by using evidence tables such as the one in Table 4.3. Free-writing—perhaps even writing out your summaries—also helps because some connections may come to you as you write. The bottom line is that writing summaries can be a useful exercise; just don't pass off the result as your final version.

Limit the Scope and Define Your Focus as a Working Title

Perhaps the most difficult aspect of writing a literature review is developing the content. Not only should a literature review be timely, but it should also differ in content from other published literature reviews—quite a challenge when the topic you have chosen has been described in many current reviews. The first thing you should do is to define the scope of your review by answering this question: what have people been studying in the past 4 years? Limiting your scope to the recent literature will not only help you write a timely survey (as opposed to a historical treatise) but also help identify the focus of your review. For example, let's say you initially wanted to write a review about a technique, Förster (fluorescence) resonance energy transfer (FRET). Your literature search, however, resulted in many more recent articles about a variation of FRET called small-molecule FRET (smFRET). Instead of writing a broad perspective of FRET (of which many already exist in the literature), your literature review would be more relevant (and perhaps more interesting to write) if you focused instead on smFRET.

Creating a working title will also help you develop your content by helping you articulate the focus of your review. This working title may be tweaked or changed as you write the review: its role is to direct the writing toward a consistent goal.

See Chapter 2 for more advice on constructing titles; however, a quick review of even the working titles in Table 4.4 and those in Chapter 2 shows characteristic differences. Titles of research articles tend to include details identifying the specific experiments discussed. In contrast, each review covers many experiments. Therefore useful titles must identify the general field and then the specific focus of the review. A good way to identify the focus of your review is to consider how you want to synthesize your information.

TABLE 4.4 Ways to Synthesize Information to Refine
the Focus of a Literature

Synthesis Strategy	Focus	Working Title
Highlight trends	Development of anti-diabetic drugs	Recent research on antidiabetic drugs
Explore or illuminate a controversy	Relationship between insulin and inflammation	Current controversies about type 2 diabetes
Evaluate the strengths and weakness of a specific approach	Sulfonylureas as anti-diabetic drug	Sulfonylureas in treating diabetes
Samples center on the general topic of diabetes.		

Structure Your Literature Review

Defining the focus and scope of your review will help you structure it. A literature review has three parts: the introduction, body, and conclusion. The purpose and estimated length of each section are listed in Table 4.5.

TABLE 4.5 Sections of a Literature Review

Section	Purpose	Length*
Introduction	Provides background to topic; defines theme and scope of literature review	One to three paragraphs
Body	Elaborates upon the scope of the review	Three to five subsections (two to four paragraphs per subsection)
Conclusion	Summarizes the current state of the field, and speculates about the future	One to three paragraphs
*Estimates are based upon a 10- to 12-page (double-spaced) literature review.		

The process of drafting the literature review is not as straightforward as that for the research article described in Chapter 2. You may find it easier to write the introduction before the body because the introduction allows you to become familiar with the history of the field. After writing the body, however, you may need to revisit the introduction to ensure that the introduction only contains information that the reader needs to know. It is difficult to capture this iterative process in the linear format of a textbook. Nevertheless, we will now describe writing these sections in the order in which they appear in a review, but keep in mind that your drafting process will involve switching back and forth between the sections.

Introduction

The introduction section provides a little background information on the topic, but more importantly, the section defines the theme (e.g., general topic) and scope of the literature review. Depending on the full length of your review, limit your introduction to one to a few paragraphs. You may find, however, that drafting longer introductions will help you process the history of the field better and identify topics that could be articulated into subheadings for the body. So all your background reading will not go to waste! The end of your introduction, like your title, should state succinctly the focus of your review. A good way to start this sentence would be something like, "This review will survey . . ."

....

In the United States, the cesarean section is one of the most commonly performed surgical procedures (Zhang et al. 2010). The procedure has become safer over time, but a cesarean section is still a surgical procedure that requires healing time for the abdominal and uterine incisions and comes

← Define the general topic early in the introduction.

with scarring and adhesions (Lavender et al. 2006). In addition, cesareans carry a greater risk of respiratory distress for the newborn upon birth and may impact a mother's future reproductive prospects because abnormal placement of the placenta is more likely to occur in a second pregnancy after a first cesarean (Lee 2008). Despite the risks of cesareans, the national rate of cesarean delivery has been on the rise since the mid-1990s (Declerq et al. 2006). The 2007 cesarean rate has more than doubled since 1996, and 31.8% of all U.S. births in 2007 were cesarean deliveries. The rate is predicted to continue to rise in the near future (Zhang et al. 2010). This literature review surveys the studies done in the last six years and suggests that non-medical, socio-economic factors may be causing the sharp increase in cesarean rates.

← Justify the literature review.

← Explicitly state the focus of the review by the end of the introduction.

Body

The body of the review is generally divided into subsections. Each subsection should adequately discuss one aspect of your topic and generally considers at least two research articles. Subheadings should clearly convey the relationship between subsection content and the focus of your review; constructing a flowchart like the one in Figure 4.1 will help you develop relevant subheadings. You can also map your ideas to see how subtopics relate to each other. The map in Figure 4.3 was drawn for a literature review on pluripotent stem cells for modeling liver disease; the map helped the student not only to structure her whole review, but also to generate the subheadings shown in Table 4.6 and to see the connections between them.

Each subsection synthesizes the literature, rather than listing summaries. Indeed, not much about a specific article is described in a professional literature review; the information

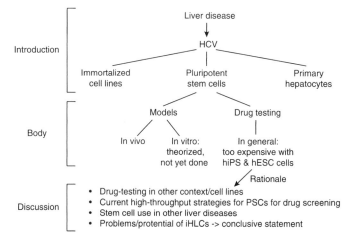

FIGURE 4.3   Mapping Ideas Can Help You See How They Relate
to Each Other.

TABLE 4.6   **Title and subheadings of literature review resonate
with each other.**

Structural Element	Example
Title states the overall topic and may specify one aspect of the topic.	Pluripotent Stem Cells (PSC): A Powerful Tool for Liver Regenerative Medicine
Subheadings should clearly relate to the overall topic conveyed in the title.	In vitro disease models to test possible therapy by PSC
	In vivo disease models to test possible therapy by PSC
	Drug and chemical enhancement of PSC differentiation in disease models
Example title and subheads are based on idea map in Figure 4.3.	

that is cited depends upon its context. To illustrate the difference between summarizing and synthesizing, let us compare
two versions of a paragraph.

Although both paragraphs cite only two research articles,
the first paragraph includes more irrelevant information about

Needs Improvement

The exceptional photostability of quantum dots (QDs) allows for long-term monitoring experiments, such as cell lineage studies. Lei and his colleagues (6) evaluated the feasibility of using QDs to label cells by introducing TAT-peptide QDs into mesenchymal stem cells, and injecting these cells into mice. The QDs nonspecifically distributed throughout the stem cells and QD-TAT peptide fluorescence was later observed in liver, lung, and spleen cells, with no (or nearly no) accumulation in the brain, heart, or kidney. This study provided a proof-of-concept for the use of QDs in cell lineage studies. Later studies showed that QDs had the long-term photostability and resistance to chemical degradation necessary for cell lineage studies. In a later study (24), QDs were conjugated to hyaluronic acid (HA) to determine how modification of HA affected its distribution in the body. Fluorescence from the HA-QDs were detectable up to 2 months after injection, demonstrating that QDs have the long-term photostability and resistance to chemical degradation necessary for cell lineage studies. Cell lineage studies like these can help developmental biologists study both the early embryo and also how stem cells are regulated, differentiated, and directed.

> Although paraphrased, the reader is practically given a summary of the paper. The purpose and methods of the research are unnecessarily described.

> Relevance of the article for the review is not stated until the end of the article summary.

Better

The exceptional photostability of quantum dots (QDs) allows for long-term monitoring experiments, such as cell lineage studies. Lei and his colleagues first provided a proof-of-concept for the use of QDs in such studies (6). Injection of mouse mesenchymal stem cells labeled with TAT-peptide conjugated QDs resulted in fluorescence in liver, lung, and spleen cells, with little fluorescence in the brain, heart,

> Relevance of the article is stated when the article is introduced, so the reader immediately understands the connection.

> The methods are described, but the description is combined with observations. Notice, too, how only some results are included.

or kidney. Later lineage studies demonstrated that QDs have the required long-
term photostability and resistance to
chemical degradation. QDs have been found
to remain active and visible in vivo for up to
two months with no ill effects on the subject
(24). Cell lineage studies like these can help
developmental biologists study both the
early embryo and also how stem cells are
regulated, differentiated, and directed.

the articles. Indeed, the first paragraph practically summarizes each article by including the motivation and implication
of the study, and a detailed description of the methods and
data. Notice, too, how the revised paragraph introduces each
article by explaining its relevance for the review, whereas the
first paragraph makes the reader wait until the end of each
summary to understand why this work is discussed.

Evidence tables can help you select the appropriate information for a particular paragraph. Take a look again at Table 4.3,
which was prepared to develop a review on the protein BCR-
ABL. These three articles could be cited in a subsection that
discusses methods to discover new mutations of BCR-ABL;
the same articles could be cited again in a different subsection
that describes the various mutations of the protein, and their
effect on protein function. See Chapter 7 for a more detailed
explanation of compressing your sources.

In general, the body of a literature review objectively surveys
the literature. Some evaluation, however, may be necessary to
highlight the strengths or weaknesses of a particular perspective. To be objective does not mean to lack an opinion—indeed,
you form an opinion just by choosing articles to include or ex-
clude. Try, however, to be fair: if you find a number of articles
that support a particular point, make sure that the articles are
from different laboratories, and consider reasons why a minor-
ity point of view is not as popular.

Animal research has provided more contradictory evidence concerning plaque formation after traumatic brain injury (TBI). Both rodent and swine models have been used extensively in Alzheimer's disease and concussion research. Both models have provided evidence of increased Aβ peptide levels post-injury (Johnson 2010). Yet, most rodent studies have not provided evidence for Alzheimer's-like plaque formation post-TBI. The rodent model does not appear to exhibit the same type of aggregation of Aβ peptides; however, accumulation of non-aggregated Aβ peptides does appear to have toxic side effects. In some cases, accumulation led to neuronal cell death, memory deficits, and behavioral problems (Smith 1998). In contrast, swine models have proved more similar to humans in their response to TBI. Swine models have shown diffuse plaque formation similar to human concussion cases, though the gross number of plaques seen in swine is far less than that observed in human brains (Smith 1997).

> Informative topic sentences clearly prepare the reader for the subject of the paragraph.

> The writer evaluates both animal models.

Conclusion

The conclusion of your review should summarize your review and speculate upon the future of the field. This is the section in which you may express your opinions most strongly, but make sure that your recommendations for the field are based on your readings and not simply pulled out of thin air. Review the Discussion section in Chapter 2 for advice on appropriate levels of hedging. In addition, the conclusion should not introduce new material or citations. Like the introduction, the conclusion should be one to three paragraphs, based on the overall length of the literature review.

Your conclusion should resonate with the focus indicated in your title and expanded in your introduction. After summarizing your focus briefly, you can then report what you have derived from the papers you analyzed in the body of the review. Discuss any predictions for the future or questions that must be answered. Be careful to clearly identify the studies on which you base these conclusions.

Many clinical milestones over the past century shaped the evolution of PCI and CABG. From Hale's brass pipes, glass tubes, and goose tracheas to Bigelow's canine cooling experiments, to Forssman's 28 self-catheterizations, the technical sophistication of therapies available to heart attack patients today reflects a long road of many trials, errors, and successes. Even today, treatments for cardiovascular disease continue to evolve. In the clinical research field, flow competition between native and grafted vessels has been postulated to cause graft failure. In the pharmaceutical arena, statin drugs have been shown to interrupt cholesterol biosynthesis in the liver, regress LDL cholesterol levels in intense doses, and induce a variety of other beneficial pleiotropic effects. How physicians can incorporate research on flow competition and aggressive statin therapy with the existing surgical options, PCI and CABG, in a way that exploits the strengths of each treatment and minimizes invasiveness for the patient, remains an area for further research. An integrative approach incorporating preexisting surgical procedures with advance in medical therapy and clinical research may provide the way for optimal revascularization.

This one-paragraph conclusion summarizes developments in the treatment of cardiovascular disease. Note the lack of new citations.

The writer highlights a future challenge for researchers.

The conclusion ends with an overall implication.

Adapt or Design Illustrations

Illustrations help readers visualize difficult concepts. Diagrams are especially useful in summarizing the current state of knowledge of a field. Review Chapter 2 for tips in designing illustrations.

In many cases, however, you do not need to design illustrations from scratch but can adapt illustrations from the articles you are reviewing. Such illustrations—especially the legend—should be modified because you should explain only those elements of the illustration that are relevant for your review. Your legend should also state whether the illustration was adapted or reproduced, and cite the original source.

Cite Sources of Adapted or Reproduced Illustrations

For every illustration that is adapted or reproduced, the legend should cite the source of the illustration. If your article will be published (e.g., online or in a thesis), you will need to get permission from the copyright holder, which is typically the journal in which the article appeared. If your request is granted, you will usually be told how the publisher wants the illustration to be cited in your work.

For our example, we will revisit a sample illustration from Chapter 2. Let us say that Figure 2.3 was not only selected for publication (congratulations!) but also chosen to be included in a literature review. In some cases, just modifying the figure legend would suffice.

Original Legend

Inhibition of TrkB function correlates to an overall decrease in PSD-95 puncta (spine) density. Neurons were treated with 1NM-PP1 (n=5, black squares) to inhibit TrkB; treatment with Bph-PP1 (n=5, gray circles) served as negative control. Individual spines budding from the dendritic processes were manually traced and counted at a given distance from the soma. ***, **, * indicate significance (P < 0.001, P < 0.01, P < 0.05, respectively) based on Student unpaired t-test; error bars=s.e.m.

PSD-95 and chemicals used to treat neurons are omitted in the literature review legend.

The number of samples is not indicated in the literature review legend.

Literature Review Legend

Chemical inhibition of TrkB leads to a decrease in the number of spines along the length of dendrites (black squares) compared to the number of spines along dendrites treated with an inactive chemical (gray circles). Figure adapted from Ref. 5.

Only the data symbols are explained.

As you can see, the legend for the literature review left out a few details like the number of samples and the chemicals used to inhibit TrkB, but it still explains the details that were retained. Such a figure could be included on a literature review on TrkB or neuron growth.

The illustration could also be stripped of many details. For example, Figure 4.4 may be included in a mini-review that advertises the original research article. Because these mini-reviews are written for a wider audience, the writer of the review needs to include the minimum number of elements necessary to convey the point—here, the decrease in dendritic density caused by inhibiting TrkB.

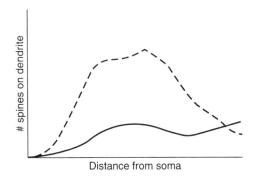

FIGURE 4.4   Inhibition of TrkB Decreases Density of Dendritic
Spines. Neurons were treated to inhibit TrkB (solid
line) or leave TrkB unaffected (dashes). (Figure
adapted from Ref. 3.)

Finalize Your Title

When you have finishing drafting your literature review, it is
time to refine your title. The biggest pitfall in creating a title for
a literature review is to make the title so general that the reader
has no idea of what is being covered. In Table 4.7 we take the
working titles that we discussed previously (Table 4.4) and
generate better titles by making them more specific.

TABLE 4.7   Revise Title for Impact

Working Title	Final Title
Recent research on anti-diabetic drugs	The diabetes pipeline: current strategies in the development of antidiabetic drugs
Current controversies about type 2 diabetes	Relationship among inflammation, insulin resistance, and type 2 diabetes: "cause or effect"?
Sulfonylureas in treating diabetes	Sulfonylureas as diabetes therapy: should it be continued?

Edit

Polishing your language helps make your literature review more professional. Because the Introduction of a research article is a type of literature review, we refer you to the Introduction section in Chapter 2 for more details on these points:

- Use correct verb tense: past tense when introducing experiments, present tense for biological facts.
- Use "we" when describing your own work, even for single-author articles.
- Format your citations and references according to the style required by the publication.

LITERATURE REVIEW CHECKLIST

Do

✓ Search for timely research articles using appropriate search engines and the Cited Literature sections of published work.
✓ Critically read and annotate the literature.
✓ Synthesize your findings around a defined focus and scope.
✓ Select a topic that is being actively studied so that you will have many points of view to synthesize.
✓ Construct a title that clearly identifies both your topic and your focus.
✓ Craft subheadings that give a solid structure to your review.

Don't

✓ Write a history of the field. Begin by concentrating on the developments of the last 4 years to find a question being studied actively by many scientists.
✓ Default to a series of summaries.

STRATEGIES FOR ORAL PRESENTATIONS AND SCIENTIFIC POSTERS

5

Have you ever heard a lecture that was so clear and interesting that you overcame your shyness and introduced yourself to the professor so that you could talk more? Have you ever viewed a TED talk on the web that inspired you not only to share the video on social media, but also to contact the presenter with a question? If so, then you can understand the power of oral communication. We both have had the experience of knowing no one at a conference until we gave a talk or presented a poster. After the presentation, we could sit down at any meal and get a good conversation going. Sometimes, we were able to make connections that would help us professionally later on.

Oral presentations and **scientific posters** contribute to research articles by providing the writer opportunities for immediate feedback. Indeed, the progress of a project is often presented orally—via either slides or posters—long before a research article is written. Even after the publication of an article, the project may be described orally. However, oral presentations are limited by time and rely more heavily on visuals. Therefore, a biologist must learn how to adapt a story to fit within the time and media constraints.

Oral Presentations

Biologists have many opportunities to present research orally (Table 5.1), such as speaking at conferences or presenting seminars as part of a job interview. Less formal presentations include those given to a biologist's research group, either about his or her own research (called a group meeting) or from a published paper (journal club presentation). Many college courses also require oral reports. You can think of an oral presentation as an adaptation of a research article, the way that a movie is sometimes an adaptation of a book. Although both the oral presentation and research article share a topic and overall structure (Introduction, Methods, Results, Discussion), an

TABLE 5.1 **Types of Oral Presentations in Biology**

Type	Content	Audience	Length
Journal club (JC)	Published research from other labs	Lab group	Depends on JC format. Some expect audience to read paper(s) and others do not. One format consists of a 10-minute presentation with a question-and-answer (Q&A) session that lasts up to 50 minutes. Another format is a presentation up to 50 minutes with interspersed Q&A.
Group meeting	Own research	Lab group	Depends on lab customs: 10–45 minutes, with Q&A up to 50 min or longer. Q&A is often interspersed with presentation.
Seminar	Own research	Biologists at speaker's own institution or other institution	45 minutes, followed by 15 minutes of Q&A
Conference talk	Own research	Conference attendees	12 to 20 minutes (depending on the conference), with 2 to 5 minutes for Q&A

oral presentation does not mean you can simply read the text of the article or use the same illustrations.

Aim for the Right Audience

The first thing you should do when preparing an oral presentation is to identify the audience. Who are they? What do they know? Will they be members of the general public or biologists? Biologists within your field or outside your field? Students or practicing scientists? Each group will arrive at your talk with different familiarity with the topic (or biology in general) and different expectations. Tailor your talk accordingly in order to keep your audience engaged.

Many of the differences between an oral presentation and a written research article stem from the differences in constraints of time. Readers of research papers theoretically have an infinite amount of time to read your paper, but the audience member of an oral presentation is attentive only for the length of your talk. Thus a reader has time to analyze your work more deeply than an audience member. In addition, a reader can control the pace and go back to an earlier section; an audience member would be considered rude to do either. To summarize, an oral presentation is the movie version of your research article: because time is limited, you need to condense your story while maintaining its integrity.

Your Audience Wants a Good Talk

Being nervous before a talk is normal. You're thinking about all the things that can go wrong and you're sure the audience will hate your presentation. But consider this. No one goes to a presentation hoping to see the speaker fail. Every person in the audience is there to learn and be inspired by you!

Plan for the Allotted Time

Most oral presentations are usually strictly time-limited. At scientific meetings, talks are organized into sessions, with each talk timed to ensure that the session is synchronized with the rest of the meeting. As you can see in Table 5.1, conference talks are limited to 12 to 20 minutes, which includes 2 to 5 minutes for a question-and-answer (Q&A) session. In contrast, job seminars and invited seminars at other institutions usually last slightly less than an hour, including Q&A. The length allows the audience members time to arrive from their last class and time to get to their next class after the seminar.

After you have decided who your audience is, you need to decide how much of your story can comfortably be told in the time you have. (Check whether your session is limited to 12, 15, or 20 minutes, because minutes count.) We find that cutting a talk after it is written is *much* harder than setting the time at the beginning. Do not make the cardinal mistake of simply speeding up your speaking rate to fit in everything you have done! Instead, decide on what part of your story can be told with a few uncrowded slides (about one per minute) plus time to explain them carefully. Also avoid the temptation to let your talk expand into the Q&A session. Questions are valuable to you because they give feedback to sharpen your presentation. You can also show your extra slides, if needed, during the Q&A.

Better yet, aim for a talk that lasts 80% to 90% of your allotted time. This advice comes from Garr Reynolds of *Presentation Zen*, inspired by the Japanese concept of eating, *Hara hachi bu* ("Eat until 80% full"). Most of your audience members will be seeing the material for the first time, so it is better to get across one point clearly and deeply than to overwhelm them with information. Planning a shorter talk will also allow you to present at a more relaxed pace (Reynolds 2011).

Craft Your Story

Many people assume that scientists are only interested in hearing the data. Facts, however, are much easier to remember when they are placed within the context of a narrative. For example, you probably remember Archimedes' principle (any floating object displaces its own weight of fluid) very well because you associate it with Archimedes jumping out of his bathtub after he realized how he can weigh the king's crown (Figure 5.1).

FIGURE 5.1   Archimedes About to Jump Out of the Bathtub.

Crafting a story for an oral presentation is very similar to shaping a research article (for more detail, see **Results: connect illustrations and text** in Chapter 2):

1. Begin where you want to end: what is the overall conclusion or argument of your work?
2. Select and organize your data into a logical story to get you to the conclusion. You may actually go back and forth between this and the previous step to solidify your story.
3. To create a proper narrative, take note of the purpose, general method, relevant data, and conclusion of each experiment.

4. Prepare an introduction, with a clear aim, justification, and context appropriate for your audience.
5. Go back to the conclusion and consider future plans or implications of your study.

Test the Structure of Your Talk with a Storyboard

Once you have determined the major content of your presentation, it is helpful to storyboard your talk. A storyboard is a sequence of sketches that help filmmakers plan their shots. Storyboarding allows you to see your talk at a glance, making it easy to organize and modify the slides as necessary. We recommend using sticky notes and a Sharpie: if your idea cannot fit on a single sticky note, then it will not fit on a single slide. The sticky notes can also help you refine the titles of your slides.

Figure 5.2 shows a storyboard of sticky notes with draft slide title. This storyboard has six types of slides found in all oral presentations. The slide estimate is based on a 10-minute talk (plus 2 minutes of Q&A).

- One title slide that states the title of your talk, your name, and your institutional affiliation.
- One or two introduction slides that explain the motivation for your project.
- One outline slide that states the overall goal of your project and illustrates your general strategy. This slide should be shown about the third slide into your talk (i.e., when your talk is one-third done) to give you plenty of time to present your data. One of the mistakes that students often make in oral presentations is to create an outline slide with this bulleted list: Introduction, Methods, Results, Discussion. This type of outline slide is not effective because biologists already expect such a structure. A stronger outline slide lists the sets of experiments that will be described, like the subheadings of the Results section of a research article (see Chapter 2). Such an

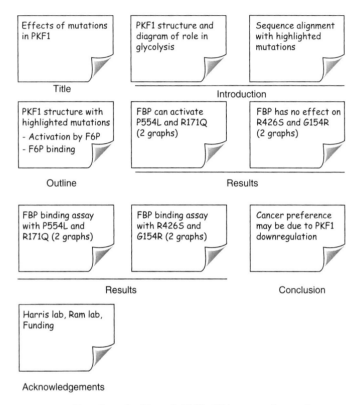

FIGURE 5.2   Storyboard of Sample Talk. Slides are to be read across, from left to right. Note that each slide has only one message, which can be modified to be the slide title.

outline slide can be shown multiple times during the presentation to help smooth the transition between sections.
- Four or five results slides that form the bulk of your presentation.
- One conclusions slide that lists your major findings.
- One acknowledgments slide that lists the people who have helped you with your research. This slide is typically shown toward the end of the presentation.

Journal Club Presentation

Journal clubs (JCs) form for different reasons: groups of people with closely related interests who want to discuss the latest work in their fields, graduate students who want to broaden their knowledge of current work for a general exam, members of a research group who want to expand their experimental program in a new direction. There are as many different JC formats as there are types of JCs.

Some classes include JC sessions for students to gain experience in reading and critiquing research articles. In those JCs, students read the paper before the JC session. The speaker begins the meeting by giving a brief (10 minutes) presentation, whose structure is similar to that of other oral presentations: introduction and general methods of the project, description of the three most important illustrations (with commentary on the appropriateness or limitations of each experiment), and implications and future directions of the project. The speaker is then expected to facilitate a discussion about the article.

Design Your Slides

For a standard slide, we recommend what educator Michael Alley calls the Assertion-Evidence slide for its efficient use of space (Figure 5.3) (Alley 2013).

The assertion provides the title for the slide. To facilitate reading, we recommend four tips:

- Write out the assertion using sentence case, i.e., capitalize only the first word.
- Use sans serif font.

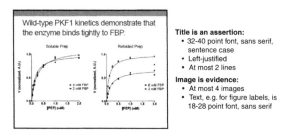

FIGURE 5.3 The Assertion-Evidence Slide Most Effectively
Conveys Your Message.

- Left-justify the title, because the upper-left corner is the first place a reader looks for information.
- Adjust the content of your assertion depending upon the type of slide. An assertion for an introduction slide could be a statement that motivates your research (e.g., "Glycolytic enzyme PGK2 is preferentially expressed and mutated in cancer"), while your final argument (prefaced with "In sum,") could begin your conclusions slide. A good title for a data slide is the conclusion of the experiment.

Evidence on an Assertion-Evidence slide typically consists of an image, so much of the advice we gave in the illustrations section of Chapter 2 would apply here as well.

Finally, notice what the Assertion-Evidence slide lacks. Other than an unobtrusive slide number, the slide lacks any labels such as an institutional logo or title of the talk. Slide numbers are useful to include because they can help an audience member refer to a particular slide during the Q&A. On the other hand, logos and the titles of a talk are unnecessary and potentially distracting: you don't introduce yourself every time you speak, so why should your slides? If you are obligated to include a logo, consider putting it only on the title and conclusions slides.

Where to Find that Image for Your Title Slide

For more generic images (e.g., for your introduction slide), consider using Creative Commons or .gov sites for high-quality, free images. The photos will be more memorable than clip art. Make sure, too, that the image has a clear relevance to the subject of the slide. Otherwise, the audience will spend more time wondering why the slide shows a snowboarder than trying to understand your graph of the enzyme kinetics.

Edit Your Slides

As you revise your slides, keep in mind the following tips.

Make the Slide Title Specific to Your Research

Instead of generic titles like "Introduction" or "Results" (terms that can title many slides), introduce your research with a complete sentence that motivates your research (e.g., "Rainforests are crucial for maintaining biodiversity and climate") and close your presentation with your overall argument prefaced by "In sum..." The most effective title for a data slide is the conclusion of the data. Because your audience will be introduced to this material for the first time, "giving away" the take-home message as the title will help audience members evaluate what you say as they examine the rest of your slide.

Keep the Text Readable

In a large lecture hall, anything smaller than 18-point font is hard to read. The advice applies especially to figure labels.

Cut Unnecessary Text

A slide deck (set of slides) should not stand on its own. If people could simply read your slide, then why would they need you? Try to make your slides enhance, not replace, your presentation by limiting bullet points to phrases (i.e., incomplete sentences) of two lines each, and avoiding secondary points (Table 5.2). If you worry about forgetting something, insert key terms in the text—particularly the title—of your slides. You can also jot thoughts down in the "Notes" section of your presentation editor. These notes are visible to you during the presentation by displaying the slides in "Presenter View" or similar mode—or as a backup, on a printed PDF of the Notes pages, with each page containing a slide and the corresponding notes. The PDF can later be sent to audience members who ask you for your slides.

Avoid Bulleted Lists
To decrease text and avoid the temptation of reading the slides, we generally recommend avoiding bulleted lists on all slides except for the outline and conclusion slides (see Table 5.2). If you must show a list, here are ways to help people remember the information better: (a) limit the list to four items; (b) attach each point to an image; (c) use animation to introduce each point. Keep the animation simple, however: elements flying in and out are unnecessarily distracting.

Keep Images Simple
Remember that your audience will be viewing your data for the first time, so simplify your slides as much as possible. Some of the methods we presented for simplifying illustrations for literature reviews in Chapter 4 apply here as well. Label samples using phrases that are easy to understand, and remove anything such as data points or trends that you will not discuss. There may be times, however, when you will want to show an image of a signal transduction or ecological pathway to demonstrate the complexity of the system. In these cases, show the pathway, but quickly zoom in to the smaller set of components on which you will focus. Otherwise, you will lose

TABLE 5.2 Slides are Better with Informative Titles and Less Text

Type	Needs Improvement	Better
Introduction: justify your research		
Outline: state your research goal		
Results: state the conclusion of the data		
Conclusion: restate your argument		

your audience in a maze of arrows and symbols, what the statistician Edward Tufte calls "chartjunk" (Tufte 1983).

Rehearse Your Presentation

Once you have your slide deck, you may feel ready to deliver your product—but not so fast! You need to rehearse your talk first (Figure 5.4). Rehearsing will help you feel more confident in your material and will help you identify aspects that need to be changed. Perhaps the most important reason to rehearse is to make sure that you are within the time limit. There are few things more annoying than a speaker who goes overtime—especially right before a coffee break or, even worse, right before you are scheduled to speak.

FIGURE 5.4 Going Through the Motions of Your Talk Helps
Create a Physical Memory of Your Presentation.

You may be tempted to memorize your talk. This strategy may be fine for very short talks (e.g., at most 3 minutes), but audience members prefer spontaneity over canned speeches. Consider, however, memorizing the first and last lines of your talk. Memorizing your first lines may help you get over the inevitable nerves. We recommend starting by introducing yourself and giving the goal of your presentation (e.g., "Good afternoon, my name is Robin Strike and today I will describe my study of APP processing and the implication for Alzheimer's disease"). Continue by characterizing the general problem (e.g., "Alzheimer's disease is . . .") instead of explaining the terms (e.g., abbreviations like APP) of your presentation title, because the definitions will make more sense as you provide background information for your project.

Memorizing the last lines will help you end on a strong note. Your ending should serve two purposes. The first is to remind the audience of your take-home message and broad impact of your study, like the last paragraph of the Discussion section of a research article, discussed in Chapter 2. The second is to let people know that you are done so that they can clap. Therefore, your very last line could be as simple as, "With that, I thank you for your time, and look forward to your questions."

As you rehearse, there are two aspects of delivery that deserve special attention. The first aspect is voice. Although you may not have the powerful voice of your favorite stage actor, you can use some of the techniques of these performers:

- Enunciate your words, and try to project your voice to the back of the room.
- Vary your tone (no one likes a monotone), but try to avoid frequent "upspeak" (a rise in intonation at the end of a sentence, making it sound like a question). Upspeak makes you sound less confident of what you are saying.
- Speak at a conversational pace and speed.
- Minimize fillers such as "like" and "um." One trick is to speak more slowly and try to catch yourself before you say "um"—and simply pause instead.

The second important aspect of delivery is "stage presence" (Figure 5.5):

- Pay attention to how you dress. You may not need to wear a suit, but at least avoid message T-shirts and caps.

- Try to stand with good posture to convey confidence in your work, and use the right amount of movements. Too many movements will be distracting, while too few make you seem stiff.
- Make eye contact with your audience; make them feel that you are talking with them, not at them. Communications expert Garr Reynolds suggests a tip from professional singers (Reynolds 2011): do not scan a room in a general way, but instead look at a few people across the room directly in the eyes (not above or below). Not only will those people sense the eye contact, but others sitting near them will feel it as well.
- Calm yourself by breathing slowly. Psychologist Les Posen suggests that you can decrease your fear with slow and deliberate breaths, taking a little more time to exhale than to inhale (Reynolds 2010).

FIGURE 5.5 Your Stage Presence Should Convey Confidence.
Be proud of your work!

Don't Forget to Practice the Q&A!

It is important not only to deliver your talk effectively, but also to conduct yourself appropriately during the Q&A following your presentation. After ending with the acknowledgments slide, put your conclusions slide up again to remind your audience of your main points. Try to anticipate questions that your audience may ask by bringing extra slides. Always paraphrase audience questions so that the entire audience hears the question. Repeating the question also gives you time to think about the question and ensure that you're interpreting it correctly. Finally, try to give concise answers, but don't try to "wing" questions if you don't know the answer. Your audience member may know more than you! It is better to admit your ignorance, and thank the questioner for suggesting a potential area of research.

ORAL PRESENTATION CHECKLIST

Do

✓ Consider your audience and craft your story before you design your slides.

✓ Make efficient use of slide space by using the Assertion-Evidence model.

✓ Rehearse your presentation.

✓ Speak at a relaxed and natural speed.

✓ Start and end your presentation on a broad note—and signal when it is appropriate to applaud.

Don't

✓ Treat your slides as documents: slides should enhance, not replace, your talk.

✓ Cram in a ton of information: instead, provide information for 80% to 90% of the talk's length.

✓ Include illustration details that you will not discuss.

✓ Use fonts smaller than 18 point, even for figure labels.

✓ Use features such as flashy slide transitions.

Scientific Posters

The poster is a visual form of communication that summarizes and advertises your research. The scientific poster may be the most frequent form of communication for scientists-in-training outside of their lab. The form encourages conversation with colleagues, providing more immediate feedback on the work and sometimes leading to collaborations.

Despite the many advantages of the scientific poster, a scientist faces a number of challenges. Scientific posters are typically presented at poster sessions, which differ significantly from the forum of an oral presentation. As you can see in Figure 5.6, the audience for scientific posters is standing, meaning that you will not have much time to discuss your work; indeed, viewers generally do not spend more than 5 minutes on a poster. In addition, there are many potential distractions in a poster session. Not only are there other posters at the session, but scientists also attend the session to chat with people they know (and to partake of the free food and drink).

FIGURE 5.6 Poster Session.

So, on the one hand, it is up to you to create a poster that can entice viewers away from other distractions. In addition, you want to make a poster that is fairly self-explanatory so that you can step away to view the other posters (or get some free food).

Prepare the Elements of Your Poster

Composing a poster is similar to creating an oral presentation: choose your illustrations, storyboard your poster, craft a narrative. Just like for an oral presentation, the illustrations you create for a poster should have fewer details than those for a research article because people spend a limited amount of time at a poster.

In addition, because your poster should focus mostly on your project, more space on the poster should be devoted to the Methods and Results sections than the Introduction and Discussion sections.

Design Your Poster

There are a couple of ways you can go about generating your poster on a computer. One strategy is to use whatever program you normally use for illustrations, print out the illustrations and text on regularly sized paper, and paste them on posterboard. The advantage of this format (Figure 5.7) is that it is easy to replace images or rearrange the components of the poster. Alternatively, if you have access to a large printer, you can use slide presentation software (e.g., Microsoft PowerPoint) as long as you remember to set the size of sheet to the same size as your poster (e.g., 48 × 72 inches). Figure 5.8 is an example of such a poster.

The overall layout of a poster is **40% illustrations, 20% text, and 40% white space**. Note that in terms of space, text should not dominate! Text is minimized because it slows down comprehension. You should also notice the importance of white space (i.e., the borders around the text and images). Such space is important to allow the eye to rest. Arrange the elements of your poster so that they follow the expected path of reading, from top to bottom, left to right.

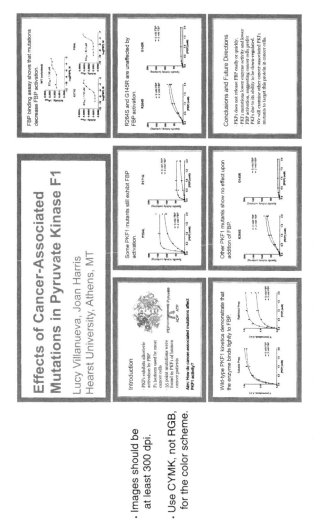

FIGURE 5.7 Poster Made Up of Individual Panels. Note the clear organization, emphasis on results, and good balance between illustrations, text, and white space.

Illustrations are vital for posters, not only to display your data but also to attract potential viewers. Review our advice on designing illustrations in Chapter 2, particularly on the minimal use of color. Figure 5.7 demonstrates more guidelines to help make your data as visual as possible on posters:

- Plot numbers of a table into a graph, at least a bar graph if no other graph is appropriate.
- Make sure the resolution of images is at least 300 dpi (dots per inch).
- Electronic displays such as computer monitors use the RGB (**r**ed, **g**reen, **b**lue) color scheme, but printers use the CMYK (**c**yan, **m**agenta, **y**ellow, and **k**ey [black]) mode. Therefore, use the CMYK format while designing your poster on the computer for more faithful reproduction of the colors.

Text is used to make the illustrations understandable and to provide context for your project. Remember that unlike a paper, your viewer will not be right next to the poster but is often about 6 feet away. Table 5.3 lists the types of text that allow readability from that distance. Note how sans serif font is recommended for short blocks of text (at most two lines) and serif font is recommended for larger blocks. That is

TABLE 5.3   **Recommended Font Characteristics on a Poster**

Type	Font Type	Font Size
Title	Sans serif	75–120 point
Names of authors, affiliated institution, email address of poster presenter	Sans serif	48–80 point
Section headings	Sans serif	36–72 point
Body (Introduction, Methods, Results, Discussion) and figure labels and legends	Serif	24–48 point
References, acknowledgments, and abstract (if necessary)	Serif	18–22 point

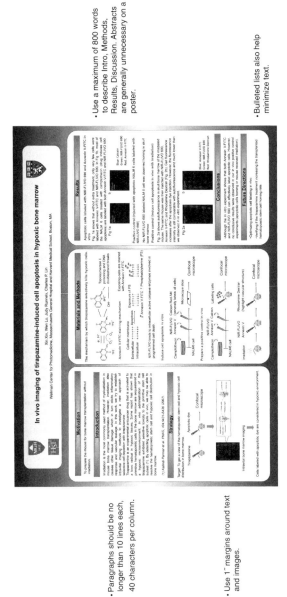

- Paragraphs should be no longer than 10 lines each, 40 characters per column.

- Use 1" margins around text and images.

- Use a maximum of 800 words to describe Intro, Methods, Results, Discussion. Abstracts are generally unnecessary on a poster.

- Bulleted lists also help minimize text.

FIGURE 5.8 Poster Printed on a Large Printer. Regardless of the format you use, keep text to a minimum.

because the serifs (little lines at the ends of each letter) help guide a reader when reading more than two lines.

Figure 5.8 shows how to minimize the amount of text on your poster, as text should constitute only 20% of your poster:

- Use a maximum of 800 words to describe all parts of your research: Introduction, Methods, Results, Discussion. We generally recommend omitting an abstract, but some conferences require that the poster includes an abstract because posters remain up for 24 hours. In these cases, the abstract can be printed in smaller type.
- Write paragraphs using a maximum of 10 lines, 40 characters per column.
- When possible, convey information in bulleted lists instead of paragraphs. Your Discussion section, for example, can simply list your conclusion(s) and impact.
- Given the importance of white space, try to make sure that your margins around the text (and images) are at least 1 inch.

Apply C.R.A.P.

The Non-Designer's Design Book by Robin Williams (the teacher, not the actor) describes four fundamental design principles that can increase the readability of your poster: contrast, repetition, alignment, and proximity (C.R.A.P.) (Williams 2014). **Contrast** is the use of visual cues to distinguish between various types of information, while **repetition** is the application of consistent visual cues to similar information. Contrast helps you attract your reader's attention, while repetition aids in creating a coherent document. The posters in Figures 5.7 and 5.8, for example, use different font sizes for the section headings versus the body text, but all the section headings within a

continued

continued

poster have the same font size. Other things that can aid contrast and repetition are font style and color.

Alignment and proximity refer to the arrangement of information on a document. **Alignment** is the intentional placement of elements such that every element is connected to another by an invisible line. **Proximity** is the clustering of related elements to create a single visual unit. Notice how the poster in Figure 5.8 separates information on the Strategy, Materials & Methods, and Future Directions, but the sections are aligned to each other along the bottom edge. Applying the principles of alignment and proximity helps create a clean and organized poster.

Edit

Before you print out your poster (especially on a large printer!), get some feedback. Even though your poster may be 48 × 52 inches, you can print the poster on regular 8.5 × 11 paper. In fact, if your text is not readable on letter-size paper, then your font may be too small. You can also receive feedback online through F1000Posters and Flickr's Pimp My Poster group.

Prepare What You Will Say

A presentation with a poster is very similar to an oral presentation, so review the earlier section on oral presentations, particularly on delivery. Here, we will highlight a few aspects on how poster presentations differ.

Because a person generally does not spend more than 5 minutes at a poster, prepare a speech that is only 3 to 5 minutes long. How do you do this? Use a one-sentence summary each for the justification, goal, overall strategy, each key result (not necessarily every single illustration!), conclusion, and impact. To create a narrative, introduce the rationale and method of each key result, and highlight the relevant aspects of the data (by relevant, we mean those data that support the conclusion of the illustration). Finally, remember as you give your speech to point to the parts of the poster that are being described.

Once you have prepared your speech, you can create a three-sentence "poster pitch." The pitch comes in handy when people come up to you and ask, "So, tell me about your poster." The pitch below is based on the poster in Figure 5.8.

Irradiation is the most commonly used technique to deplete the bone marrow of cells in mice in preparation for transplantation, but irradiation can lead to inflammation and vascular leakage. Here, we present our progress on an intravital imaging method to visualize cells ablated by an experimental cancer drug. Our work contributes to efforts to prepare mice for bone marrow transplantation without irradiation.

⟵ First sentence sets up the problem.

⟵ Second sentence summarizes the main point of your research.

⟵ Third sentence describes a broad implication of your work.

Finally, like for any oral presentation, spend some time thinking about questions you might be asked during or after your presentation!

POSTER PRESENTATION CHECKLIST

Do

✓ Print out a draft on regular letter-size paper to check for font size and to facilitate feedback.

✓ Follow a 40:40:20 ratio for illustrations:text:white space.

✓ Prepare a 3- to 5-minute speech.

✓ Use fonts larger than 18 point for the references, acknowledgements, and abstract and fonts larger than 24 point for all other text, including figure labels.

Don't

✓ Use tables or large blocks of text: instead, convert tables into graphs, and blocks of text into bullet points when appropriate.

STYLE

6

Just as biologists expect to find certain information in predictable sections of an IMRaD research article, they appreciate crisp, clear prose that strikes the right scientific register. Your style should reflect the values of science, especially precision, and account for how scientists typically read—quickly and strategically.

In this chapter we offer tips to improve readability. We also help you recognize and address grammar, usage, and style errors that we see all too frequently in student (and professional) papers, as every misstep on a page hurts your credibility with readers.

Do not be surprised if you have seen some of our advice in other writing courses or handbooks: certain general principles for style apply regardless of topic or genre. However, many of the strategies we offer are specific to writing in the sciences.

Most experienced writers defer editing for style and correctness until the final stage of the writing process. And rather than edit on a screen, most feel the need to print a hard copy and scrutinize the text with a pencil in hand—often more than once, with a different set of issues addressed each time. Achieving a clear, efficient scientific style is neither quick nor easy. What English writer Thomas Hood said about literary writing applies just as much to technical communication: "the easiest reading is damned hard writing" (Hood 1837).

Use Strong, Precise Topic Sentences

Well-crafted topic sentences orient readers to what follows. They also permit readers to scan efficiently across a section to grasp the overall flow of ideas.

Insisting on topic sentences may to some seem like outdated advice—in fact, early research in writing studies debunked the traditional "rule" about putting topic sentences at the front of each paragraph by showing that published writers of expository writing frequently omit topic sentences or do not place them first (Braddock 1974). Yet with scientific prose, announcing your key information at the front of each paragraph is wise because doing so makes reading more efficient.

Needs Improvement

Multiple strategies have been pursued to reactivate wild-type p53 function in cancer cells. One of these strategies is the delivery of wild-type copy of p53 via a retroviral infection (Roth et al., 1996). Although this type of therapy leads to a regression of the treated tumors and shows low toxicity for the normal tissue, it is hard to apply. The delivery of the virus involves an invasive procedure, which often causes complications in the patients. Another disadvantage of the approach is the low infection efficiency. Because not all tumor cells obtain a wild-type copy of p53, the surviving cells can cause the regrowth of the tumor. The efficiency of this therapeutic strategy is further diminished by the usage of a virus, because the viral construct can trigger an immune response that would deplete the p53-carrying vector before infection has occurred.

The opening sentence here is adequate but not optimal because the paragraph does not just describe multiple strategies but also (and more importantly) emphasizes their disadvantages.

Better

The use of retroviral infection to reactivate ← wild-type p53 function in cancer cells (Roth et al., 1996) has a number of drawbacks. The delivery of the virus involves . . .

> This is a more comprehensive and precise topic sentence because it signals the main purpose (i.e., to announce the drawbacks).

Don't waste the opening of a paragraph by including too much background information, detailing the chronology of methods that led up your main claim, or hedging: instead, announce the main content *and* purpose of the paragraph in a single sentence, and with confidence.

Also note that even though the topic sentence comes first, we usually cannot write a good one until *after* we write the paragraph—that is, until we know what we are trying to say.

Within a Paragraph, Employ a Consistent Subject Across Most Sentences

Topic sentences are most effective when a paragraph is coherent. In a coherent paragraph, the subjects of the sentences form a relatively small set of related topics. Compare the two passages below, in which we **highlighted the subjects**.

Needs Improvement

The **creation** of the postsynaptic density (PSD), a dense network of proteins and receptors recruited to the postsynaptic membrane in response to synapse formation, is one of the main factors affecting synaptic plasticity. In particular, **postsynaptic density-95 (PSD-95)** is a specific PSD membrane-associated protein that anchors N-methyl-D-aspartate receptors (NMDARs), which are glutamate receptors at excitatory synapses, into the postsynaptic membrane.

> The grammar is fine, but the writer does not do much to help the reader identify the topic of the paragraph.

An **increase** in NMDAR-dependent long-term potentiation (LTP), a synaptic plasticity mechanism that improves long-term memory (Stein et al., 2003), is thought to be stimulated by PSD-95 overexpression. As a result, **studies** of various memory and learning-related disorders, such as Alzheimer's disease and autism, have focused on PSD-95.

Better

One of the main factors affecting synaptic plasticity is the creation of the postsynaptic density (PSD), a dense network of proteins and receptors recruited to the postsynaptic membrane in response to synapse formation. In particular, **postsynaptic density-95 (PSD-95)** is a specific PSD membrane-associated protein that anchors N-methyl-D-aspartate receptors (NMDARs), which are glutamate receptors at excitatory synapses, into the postsynaptic membrane. **PSD-95 overexpression** has been implicated in stimulating an increase in the NMDAR-dependent long-term potentiation (LTP), a synaptic plasticity mechanism that improves long-term memory (Stein et al., 2003). As a result, **PSD-95** has become a target of study to deduce the mechanisms of various memory and learning-related disorders, such as Alzheimer's disease and autism.

This paragraph is more coherent because the boldface sentence subjects all refer to a factor called PSD-95, which helps orient the reader. Notice, however, that some variety in phrasing avoids a sense of rote repetition.

Begin with Information that is Familiar to Your Readers; Then Introduce New and Complex Information

Compare the readability of these two sentences below:

Efficient, faithful replication of DNA during amplification of recombinant plasmids for further study is something that is valued by biologists.

> Biologists value efficient, faithful replication of DNA during amplification of recombinant plasmids for further study.

The second is easier to read because it begins with something familiar ("Biologists") rather than with a long, abstract concept. In "The Science of Scientific Writing" (which you can find on the web—it is worth a look), George Gopen and Judith Swan, drawing on the work of Joseph Williams, write, "In our experience, the misplacement of old and new information turns out to be the No. 1 problem in American professional writing today." (Gopen and Swan 1990)

In general, sentences (and paragraphs) flow better when they start with information that is already known to the reader. Begin more familiar or simpler information and end with the newer and more complex information.

Let's revisit those two sample sentences above:

Efficient, faithful replication of DNA during amplification of recombinant plasmids for further study is something that is valued by biologists. ⟵ This writer puts the most abstract and complex element first, which makes the sentence hard to read.

Biologists value efficient, faithful replication of DNA during amplification of recombinant plasmids for further study. ⟵ "Biologists" is the more familiar, "old" information, and it helps orient readers to what follows.

Link Sentences when the Prose Feels Choppy or Disconnected

The same familiar-to-new principle you can apply *within* sentences can be applied to *sequencing* multiple sentences. To improve flow, link the end of one sentence to the beginning of the next. That is, the new, old, or familiar information at the *end* of the first sentence *becomes,* or at least *echoes,* the information at the *front* of the next sentence.

> Familiar element A → new/complex element B. B [linked to/echoed at front of sentence, rather than placed later in sentence] → new/complex element C.

Needs Improvement

Many high-activity DNA polymerases share an ability to proofread and correct errors in DNA replication, although fidelity rates may vary substantially. A 3' to 5' exonuclease identifies and removes mismatched nucleotides. As replication continues, proper nucleotides are inserted to replace the incorrect nucleotides.

← This paragraph is grammatically fine, but it ignores the familiar-to-new strategy at the sentence level and it makes no effort to link sentences.

Better

Although speed and fidelity rates vary substantially among DNA polymerases, a trait shared by many high-activity DNA polymerases is the ability to proofread and **correct errors in DNA replication. Error correction** is performed by a 3' to 5' exonuclease, which identifies and removes **mismatched nucleotides from DNA**. These **incorrect nucleotides** are then replaced by proper nucleotides as replication continues.

← The "new" information in bold both links to and becomes the familiar information in highlighting with the sentence that follows.

The following tips generally work well in supporting these same principles:

- Keep the subject short and specific, and introduce it within the first seven or eight words so that the reader can easily identify it.

Needs Improvement

The transfer of the phosphoryl group of phosphoenolpyruvate onto ADP, which is the final step of glycolysis, is catalyzed by pyruvate kinase.

← This complex subject makes it difficult for the reader to identify the sentence topic.

Better

The **final step of glycolysis**, which is the transfer of the phosphoryl group of phosphoenolpyruvate onto ADP, is catalyzed by pyruvate kinase.

← Simplified topic is easier to grasp.

- Keep the subject close to its verb.

Needs Improvement

Methods to identify small molecules able to reactivate mutant p53 proteins in tumor cells have also been **developed**.

← The reader may forget who is doing the action by the time she has reached the verb.

Better

Methods have also been **developed** to identify small molecules able to reactivate mutant p53 proteins in tumor cells.

← Notice how "developed" is moved closer to the subject, "Methods."

- Use the subject to convey the actual topic of the sentence, or what the rest of the sentence is about.

Needs Improvement

Structural p53 mutants have **mutations** that interfere with the normal folding of the protein.

← Real topic of sentence is the direct object.

Better

In the p53 protein, structural **mutations** interfere with the normal folding of the protein.

← Topic is more appropriately placed as the subject.

- Trim the end to clarify the emphasized information.

Needs Improvement

Completion of the final step usually involves the conscription of normal cells that reside in, or are recruited to, the tumor stroma, a **process** that marks the successful colonization of a new tissue niche that is **advantageous** to the disseminated tumor cells, due to the increased nutrient availability and relaxed spatial constraints (Fidler, 2003).

← Is the emphasis of the sentence on the process or the advantage?

Better

Completion of the final step usually involves the conscription of normal cells that reside in, or are recruited to, the tumor stroma. This process marks the successful colonization of a new tissue niche that is advantageous to the disseminated tumor cells, due to the increased nutrient availability and relaxed spatial constraints (Fidler, 2003).

← Splitting up the sentence clarifies the take-home message of each sentence.

Use Precise and Purposeful Transitional Phrases

Adhering to the old-to-new principle is sometimes not possible or creates unwanted redundancy. Thankfully, transitional words and phrases such as *However, In addition, In contrast,* and *First . . . Second . . . Finally* can also improve paragraph cohesion:

One significant limitation of in vitro and in vivo studies is their reliance on established human cell lines and xenograft studies. These approaches are unable to fully recapitulate the complexity of clinical carcinomas for a variety of reasons (Valastyan et al., 2009). **For example,** cell lines accumulate genetic mutations over multiple passages in culture. Some of these mutations may influence tumor development or metastasis in unknown ways, making it difficult to distinguish the true effects of our specific genetic perturbation (Valastyan et al., 2009). **On the other hand,** xenograft studies in mouse models fail to capture species-specific interactions that occur between the human tumor cells and the surrounding murine stromal cells. These interactions may be of relevance, particularly since successful tissue colonization and establishment of macroscopic metastases usually involve the conscription of normal cells that reside in or are recruited to the surrounding stroma (Hanahan and Weinberg, 2000).

> "For example" prepares the reader for one of the reasons without repeating the word "reasons."

> "On the other hand" signals a second limitation of previous studies.

Transitional words should be used judiciously. A paragraph with multiple instances of *However* or *Yet* disorients your reader. Similarly, some writers get stuck in a rut using one or two stock transitional words, such as *Therefore* or *For example*. Be aware of the wide repertoire of transitional phrases available to you and select the one right for the *specific purpose* that it needs to perform.

- **To add, agree:** and, also, as well as, in addition, moreover, equally, coupled with, together with, furthermore, besides, equally important

- **To signal time or sequence:** after, before, earlier, eventually, in the meantime, at the same time, simultaneously, next, later, until, prior to, since, occasionally, gradually, first/second/third
- **To introduce examples or details**: for example, for instance, to illustrate, especially, specifically, in other words, to clarify, such as
- **To signal contrast:** however, although, whereas, in contrast, on the contrary, on the other hand, nevertheless, conversely
- **To signal similarity:** like, likewise, similarly, correspondingly
- **To suggest causality:** therefore, as a result, consequently, for this reason, hence, thus, accordingly, provide that, given that
- **To show emphasis:** clearly, above all, in fact, in short, chiefly, especially, as has been noted
- **To summarize or conclude:** after all, in summary, finally, in conclusion, on the whole, to sum up, ultimately, as can be seen, overall

Choose Your Verbs Carefully: Active Versus Passive Voice

Some claim that scientists should always use the passive voice to maintain a focus on the work rather than on the researcher, but that is not what the best writers in our field do because long strings of passively voiced sentences usually hurt readability. When choosing which kinds of verbs to employ, consider the context, purpose, and audience. For example, passive voice usually is better for the Methods section to avoid the childlike tone of the active voice ("I did *X*, then did *Y*"). On the other hand, first-person, active voice is typically more effective in the Introduction and Discussion sections to distinguish your work from those of others ("We discovered . . ." rather than "It was discovered that . . ."). Active voice can also improve clarity in the Results section, but biologists rarely use the first person here; instead, the subject is more likely to be an organism, molecule, or observation:

- The Rel transcription factor **activates** . . .
- The evolutionary conservation of the viral DNA sequence **suggests** . . .

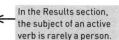 In the Results section, the subject of an active verb is rarely a person.

You can use the passive voice effectively to emphasize the appropriate information in a sentence or improve cohesion between two sentences. In the pair of sentences below, a reader would expect more information about fluorescent particles to follow the first sentence but more about cells to follow the second sentence:

- Active voice: The cells **endocytosed** the fluorescent particles. The particles. . . .
- Passive voice: The fluorescent particles **were endocytosed** by the cells. These cells . . .

Change from active to passive voice alters the expected topic of the subsequent sentence.

Regardless of which voice you use, remember that verbs drive a narrative. Therefore, the true action of a sentence should be conveyed in the verb—and not in the noun or adjective form of the verb, which is called a nominalization. Converting nominalizations into verbs often clarifies the meaning of the sentence and makes it more concise as a bonus.

Needs Improvement

Inhibition of transcription **is strongly dependent** on ATP.

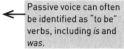 Passive voice can often be identified as "to be" verbs, including *is* and *was*.

Better

Inhibition of transcription **depends strongly** on ATP.

Needs Improvement

An **analysis** was performed to determine the function of the Ypel4 gene in fruit flies.

Better

We **analyzed** the function of the Ypel4 gene in fruit flies.

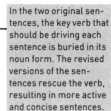

In the two original sentences, the key verb that should be driving each sentence is buried in its noun form. The revised versions of the sentences rescue the verb, resulting in more active and concise sentences.

Be Concise

Many students think that when it comes to writing papers, longer is better—indeed, they often stretch out a paper to meet a page limit or include every last living detail of what they have done because they fear not getting credit for it. However, practicing biologists (and the best student writers) put a premium on conveying information using the fewest words. This push for efficiency is often driven by the word-count limits imposed by journals and funding agencies, but the more fundamental principle is that overwriting slows down the reader without improving content or accuracy—a wasteful, lose–lose situation.

This does not mean that shorter is *always* better, because writers must balance the impulse to economize with the need to include enough detail. When pruning your prose tips over into oversimplifying, you've gone too far. Still, from what we have seen in both student and professional prose, most writers need to work on concision.

Begin at the paragraph level by eliminating anything that is not directly relevant to the topic sentence. Notice how whole sentences in the following paragraph can (and should) go:

> Erythropoiesis, or red blood cell development, is an essential process for any animal that has a circulatory system. ~~Examples of animals that lack a circulatory system are jellyfish and planaria~~. In humans, approximately 2 million reticulocytes, the final precursor to mature red cells, are produced in the bone marrow every second, and production can increase up to 20-fold under severe hypoxic stress (An & Mohandas, 2011). ~~By the end of the day, a human generates nearly 1% of his body's red blood cells.~~ Through the course of their development, mammalian erythrocytes lose their nuclei and undergo a substantial shape change before becoming functional erythrocytes circulating the bloodstream. ~~This process differs in people who have the condition known as sickle-cell disease.~~ The terminal steps in differentiation of mammalian red cells are the developmentally important steps that give rise to the unique phenotype and function of erythrocytes.

In the following passage, all the sentences are relevant, but most are bloated with unnecessary phrases:

....

Needs Improvement

Since their discovery by the scientific community ~~of microbiologists~~, bacteriophages have been seen as potential ~~and potent~~ antibiotics. Although bacteriophages were ~~relatively~~ forgotten in favor of chemical antibiotics, the ~~evolution and~~ propagation of

← Avoid this kind of filler and bluster. Some writers think it makes them sound smart, but it irks most readers.

multi-drug resistant bacteria has ~~brought about a resurgence of~~ increased interest in using bacteriophages ~~as antibiotics or as a~~ supplement to existing antibiotics. Cur-rent~~ly, there is much~~ research in the field of phage therapy~~, with scientists striving to~~ clarify the safety and efficacy of phages in antibacterial roles, as well as ~~to~~ identify the best methods of administration.

Better

Since their discovery by the scientific com-munity, bacteriophages have been seen as potential antibiotics. Although bacterio-phages were forgotten in favor of chemical antibiotics, the propagation of multi-drug resistant bacteria has increased interest in using bacteriophages to replace or supple-ment existing antibiotics. Current research in the field of phage therapy includes clari-fying the safety and efficacy of phages in antibacterial roles, as well as identifying the best methods of administration.

⟵ After cutting out filler, you often need to tweak the wording, transi-tions, and syntax.

Here are four specific and durable strategies you can apply systematically to make your prose more concise without sacri-ficing meaning:

Omit the Following Types of Words

- Empty modifiers, especially adverbs: actually, kind of, really, etc.
- Doubled words: first and foremost, true and accurate, basic and fundamental, etc.
- Words and phrases that can be inferred: ~~future~~ plans, large ~~in size~~, red ~~in color~~, results ~~that we discovered~~, methods ~~we used~~ included, etc.

- Sentence beginnings that do not contain the actual topic: *It is interesting to note that, Research has shown that, The fact of the matter is that,* etc.

Replace a Phrase with a Word

Table 6.1 lists common phrases that could be shortened to a word. You can probably think of more.

TABLE 6.1. **Common Phrases that Could be Shortened**

Wordy	Concise
the reason for	why
despite the fact	although
a decrease/increase in the number	fewer/more
more often than not	usually
due to the fact that	because
it may be that	perhaps
an additional piece of	further
in the present study	here
firstly . . secondly . . .	first . . . second . . .

Refer to Illustrations by Number, Not Location

Citing illustrations by number eliminates the need to include words that signal their location (e.g., below, above). This not only improves concision but also accounts for when, in a publication or presentation, the illustration is not placed adjacent to where you are discussing it in the text.

NEEDS IMPROVEMENT: Figure 3 below shows an increase in expression, suggesting . . .

BETTER: Figure 3 shows an increase in expression, suggesting . . .

Define a Term or Acronym the First Time You Use It; Then Use the Acronym

Note that acronyms like DNA and ATP do not need definitions because they are commonly understood.

Needs Improvement

Recent studies have also implicated a role for **fatty acid oxidation** at the start of cancer. **Fatty acid oxidation** is reactivated in cells that experience **loss of attachment** from the extracellular matrix (Schafer et al., 2009). Cells undergoing **loss of attachment** require more ATP, so reactivation of **fatty acid oxidation** may provide the additional ATP necessary to prevent apoptosis induced by **loss of attachment**.

Better

Recent studies have also implicated a role for fatty acid oxidation (**FAO**) at the start of cancer. **FAO** is reactivated in cells that experience loss of attachment (**LOA**) from the extracellular matrix (Schafer et al., 2009). Cells undergoing **LOA** require more ATP, so **FAO** reactivation may provide the additional ATP necessary to prevent **LOA**-induced apoptosis.

← Define the acronym with the first usage.

Use acronyms only when readers need them. The example is from a paper that refers many times to fatty acid oxidation and loss of attachment, so it is efficient for both the writer and reader to use those acronyms. If, however, either FAO or LOA was mentioned only a couple of times in the whole paper, introducing those acronyms would only distract the reader.

Check and Clarify Your Antecedents

Perhaps the most frequent example of imprecise language we see is incorrect or unclear antecedent. Sometimes a writer will employ a pronoun (most often *it* or *this*), but what that pronoun refers back to is ambiguous. The intended meaning may still be clear to most readers, but sometimes the ambiguity creates real confusion—and for readers who value precision, unclear pronoun references are always irksome.

Modifiers and pronouns should not only refer to the correct antecedent but also agree with the antecedent in number. Here are the most common problems:

- Pronoun does not identify its antecedent.

Needs Improvement

Our data suggest that the purified protein was already bound by the activator. **This** made it difficult to see the effect of adding activator to the reaction.

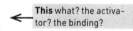
This what? the activator? the binding?

Better

Our data suggest that the purified protein was already bound by the activator. **This binding** made it difficult to see the effect of adding activator to the reaction.

Needs Improvement

Since all six bacterial strains produce high levels of inviable spores, **it allows** us to study this phenotype in different genetic backgrounds.

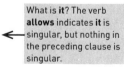
What is **it**? The verb **allows** indicates **it** is singular, but nothing in the preceding clause is singular.

Better

Since all six bacterial strains produce high levels of inviable spores, **these strains allow** us to study this phenotype in different genetic backgrounds.

- Misplaced modifier alters meaning of sentence.

Needs Improvement

Activating the PI3K/Akt pathway promotes PSD-95 transfer from the ER to the Golgi apparatus in the cell body, **which induces** a rapid widespread trafficking of PSD-95 into synapses.

What is inducing the trafficking, the cell body or the PSD-95 transfer?

Better

Activating the PI3K/Akt pathway promotes PSD-95 transfer from the ER to the Golgi apparatus in the cell body. The transfer induces a rapid widespread trafficking of PSD-95 into synapses.

- Dangling modifier confuses the implied subject of the modifying phrase with the subject of the main clause. Dangling modifiers often crop up when you use the passive voice!

Needs Improvement

After injecting the nanoparticles, the mice were placed in their cages for further observation.

This sentence suggests that the mice, not the experimenter, injected the nanoparticles.

Better

After nanoparticles were injected into the
mice, the mice were placed in their cages
for further observation.

....

This problem can even appear in titles. Consider the case of
a literature review written by a student interested in what studies were suggesting about the therapeutic uses of horseback
riding for children diagnosed with cerebral palsy:

....

Effects of Horseback Riding on Children
with Cerebral Palsy

....

This title can be read two ways: the intended meaning
(how horseback riding therapy affects children with cerebral palsy) or the unintended (and horrifying) one (how
riding horseback *on top of children with cerebral palsy* has certain effects).

A Different Pronoun Issue: Inclusive Language

You may have noticed that when we needed to employ a singular pronoun in this book, we alternated between using
"he" and "she." Using "he" as the exclusive default for a generic singular pronoun is no longer acceptable in academic
or popular prose. Some writers use "he or she" instead,
which is gender-inclusive. Those who find "he or she" clumsy
revise their sentences to make the items in them plural,
which then allows the gender-neutral pronoun "they" or
"their." That is the case with that last sentence, as well as this
one: "Many students don't know what their options are
when it comes to being gender-inclusive in their writing."

Avoid Compound Nouns

Students are often tempted to use a string of nouns (e.g., "NMDAR-dependent LTP synaptic formation events"). This tends to obscure rather than clarify ideas, especially for a reader who is not familiar with such phrases. In addition, such phrases may not be easy to decipher because they can obscure which term is modifying what. Compound nouns can be even more deadly at the beginning of a sentence because they violate the familiar-to-new and keep-subjects-simple-when-possible principles.

Needs Improvement

Kovalchuk et al. (2002) showed that this developmental pathway is also involved in stimulating **NMDAR-dependent LTP synaptic formation events**.

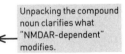

Which term is "NMDAR-dependent" modifying, LTP or events?

Better

Kovalchuk et al. (2002) showed that this developmental pathway is also involved in stimulating **synaptic formation during NMDAR-dependent LTP**.

Unpacking the compound noun clarifies what "NMDAR-dependent" modifies.

Use Technical Terms, but Avoid Jargon

Jargon is convoluted prose that may be fine for talking with your colleagues, but it is not professional enough for publication. Technical terms, on the other hand, can be efficient in helping reading comprehension, especially if you take the time to define your terms in ways appropriate for your audience.

Needs Improvement

Expression of the transgenic protein was verified by **Western blots** of proteins from the transfected cells. ← The term "Western blot" is too colloquial for a professional article.

Better

Expression of the transgenic protein was verified by **immunoblots** of proteins from the transfected cells.

Edit for Correctness

So far we have been discussing style, which traffics in matters of "more effective" and "less effective" prose. But there are also matters of grammar, usage, and word choice, which are simply "right" or "wrong" according to the standards of formal edited English. We cannot cover everything here—not even close—but we can share the items that our experience tells us surface most frequently or that scientific readers find most irksome.

- Use the Singular with None and Any

  None of the cell extracts **exhibit** enzyme activity.
  BETTER: None of the cell extracts **exhibits** enzyme activity.

Use a Hyphen to Combine Two Terms that Together Form an Adjective

The **ATP activated** protease bound to its substrate more strongly.
BETTER: The **ATP-activated** protease bound to its substrate more strongly.

We observed a **fifteen fold** increase in activity.
BETTER: We observed a **fifteen-fold** increase in activity.

Use Appropriate Nomenclature and Formatting

This is sometimes more easily said than done, especially when referring to genetic nomenclature, so you should consult the database for the organism. Italicize binary names of organisms (*genus species*), but when using only the genus, don't italicize it. After the first use of a binary name, the genus name can be replaced by an initial unless this abbreviation creates confusion with another organism of a different genus that appears in the same paper. Note the appropriate use of italics in the following example:

- The toxins of Pseudomonas aeruginosa may differ from those used by other Pseudomonas species.

Avoid Starting Sentences with Numerals

20 ml His-Trap FF Ni sepharose beads were used to purify the protein.
BETTER: His-Trap FF Ni sepharose beads (20 ml) were used to purify the protein.

Numbers less than 10 should be written out

BETTER
Nine samples were analyzed.

Include a Comma After the Introductory Element of a Sentence

This is the most widely made grammar mistake among college writers. In the popular press and in some academic fields, this rule is softening, and writers can get away without a comma after certain short introductory elements, but in most formal prose, the rule holds. If/then sentences always require a comma.

Needs Improvement

During the epithelial-mesenchymal transition epithelial cells gain mesenchymal traits.

Better

During the epithelial-mesenchymal transition, epithelial cells gain mesenchymal traits.

Needs Improvement

If cultured Drosophila cells are moved to
37°C the heat shock response immediately
reprograms their RNA transcription and
protein synthesis.

Better

If cultured Drosophila cells are moved to
37°C, the heat shock response immediately
reprograms their RNA transcription and
protein synthesis.

Avoid Ending Sentences with a Preposition

ChIP was used to identify the gene segments that the
transcription factor binds to.
BETTER: ChIP was used to identify the gene segments
to which the transcription factor binds.

Recognize the Singular Versus Plural Forms of Irregular Terms

There are a number of irregular words that do not follow the
rules for forming plurals (Table 6.2).

TABLE 6.2 **Singular and Plural Forms of Irregular Words**

Singular	Plural
alga	algae
analysis	analyses
bacterium	bacteria
basis	bases
datum (rare), data point	data
focus	foci
formula	formulae/formulas
fungus	fungi, funguses
genus	genera
hypothesis	hypotheses
index	indices (numbers), indexes (in books)
larva	larvae
locus	loci
medium	media, mediums
ovum	ova
phenomenon	phenomena, phenomenons
protozoan	protozoa, protozoans
serum	sera, serums
species	species
spectrum	spectra
stimulus	stimuli
taxon	taxa
villus	villi

Be Careful of Common Mistakes in Word Choice

- affect: (verb) to influence or produce an effect.
- effect: (noun) result, outcome; (verb) to bring about.

> We analyzed how a mutation at H147 **affects** the fidelity of *Pfu* polymerase.
>
> We analyzed the **effect** of a mutation at H147 on the fidelity of *Pfu* polymerase.
>
> A mutation at H147 **effects** a major change in the fidelity of *Pfu* polymerase.

- fewer: refers to something that can be counted.
- less: refers to something that can be measured, but not counted.

When compared to the wild-type enzyme, the mutant polymerase made **fewer** mistakes but also synthesized **less** DNA.

- its: the possessive form of *it*.
- it's: a contraction of *it is*. Contractions may be too informal for professional writing, so consider avoiding them.

Needs Improvement
It's clear that ATP released the enzyme from RNA.

Better
It is clear that ATP released the enzyme from RNA.

- **Use "that" for a restrictive or defining clause,** information that limits the meaning of a word or phrase. Such clauses tend to be introduced by **that** and are never preceded by a comma, e.g. "DNA polymerases are enzymes that catalyze the replication of DNA."
- **Use "which" for a nonrestrictive or nondefining clause,** a clause that adds information. Such clauses are preceded by a comma, e.g. "DNA polymerases, which are enzymes, catalyze the replication of DNA."

- DNA polymerases are enzymes that catalyze the replication of DNA.
- DNA polymerases, which are enzymes, catalyze the replication of DNA.

Needs Improvement

DNA polymerases, which have an exonuclease activity, exhibit high fidelity.

← This sentence suggests that all DNA polymerases have an exonuclease, which is not true.

Better

DNA polymerases that have an exonuclease activity exhibit high fidelity.

Here is a quick test to help you discern whether **that** or **which** is appropriate. If you remove the clause and the sentence still reads okay and the meaning is substantially the same, use **which** along with a comma or set of commas. If removing the clause does not allow you to complete a full sentence or fundamentally changes the meaning of it, use **that** and no comma(s).

Writing With an Accent

We would also like to recognize that many writers who use English as a second or additional language write "with an accent." Their "accented" prose may include using incorrect verb endings, articles, pluralization, and/ or prepositions. While many readers are able to read past accent, some errors may lead to the misinterpretation or misrepresentation of data. If you write with an accent, we recommend that you pay close attention to reader feedback to figure out which mistakes create problems with communication, so that you can self-edit for such errors. For high-stakes writing, such as research articles written for publication, we recommend seeking the services of an editor or native English speaker.

STYLE CHECKLIST

Do

✓ Edit your draft by looking at a hard copy for one style item at a time.

✓ Ensure that each unit of discourse (e.g., section, paragraph, sentence) has only one topic and presents ideas from familiar to new.

✓ Make the effort to do a second round of editing, even when you think you are done.

Don't

✓ Address lower-order concerns (e.g., style) before higher-order concerns (e.g., analysis, organization, paragraph-level revisions). This can slow down writing too much.

✓ Sacrifice precision for concision.

SOURCES

7

Sources are important in scientific writing because scientists never work in an intellectual vacuum: they build upon each other's work, and readers need to know the foundations of each new study. Sources are also important to provide support for specific facts or opinions stated by the author. For both these reasons, you are obliged to cite the sources of information—publications, data, images, and ideas—that you did not create. Citations allow readers to consider sources when assessing the accuracy of the information, evaluating the quality of the argument, and tracing the accumulation of knowledge over time.

As an example of how the accumulation of knowledge depends on scientists building on each other's work, let's take another look at the intellectual connections between Avery, Watson, and Crick.

At the time the Avery et al. paper on DNA was published (Avery et al. 1944), studies on plants, animals, and bacteria had led to general agreement that genetic material (chromatin) was composed of DNA and protein. Yet there were sharp disagreements about what actually carried the information. DNA was abundant and ubiquitous but appeared to be a monotonous repeat of four deoxyribonucleotides. Many people did not believe that such a repetitive molecule could carry the complex information needed to encode an organism. These people thought that protein, with 20 amino acids, was a much better candidate, while DNA must be a structural component. So, the finding of Avery and his colleagues did not fall into fertile soil.

Some people, however, did welcome the idea of DNA transmitting genetic material, and it is interesting to see how these people form prominent links in the chain leading to the double helix. Salvador Luria, a geneticist who turned to phage to investigate the nature of the gene, directed Watson's PhD thesis and later facilitated Watson's study with Crick in England. The biochemist Erwin Chargaff showed that DNA was not as repetitive as it appeared initially (Chargaff et al. 1951, 1952): organisms had characteristic base compositions that differed from species to species. Despite this variation, the %A always equaled %T, while %C equaled %G. Chargaff's rules led Watson and Crick to place A+T and G+C pairs inside the double helix when they built the model. Every biologist works within an ongoing conversation of ideas, and citations make that conversation both possible and visible.

Students tend to think of citing sources as a way to avoid plagiarism, but the story of the double helix shows that citations are necessary for tracking the intellectual development of an idea. Knowing the prevailing hypothesis and being aware of the techniques available at the time an experiment was done help explain why the experiment was designed as it was, and why today's state-of-the-art method was not used. Understanding the larger conversation of ideas as revealed in a chain of citations allows a deeper understanding of the science; such understanding may also give you ideas for developing new studies.

In Chapter 4, we discussed how to evaluate the quality of sources for your literature review. Here, we show how using your sources correctly involves three actions:

- Paraphrasing the information in the document
- Citing the source within the document as evidence: this means inserting an in-text reference every time you use a source

- Listing the reference at the end of the document so that readers can locate sources for themselves

Unlike other chapters of our book, this chapter does not incorporate examples of student writing. Instead, we illustrate aspects of referencing by highlighting important papers and books in biology. We hope these examples will not only help you use your sources appropriately but also inspire your own experiments.

You probably already highlight important points as you read, but to really understand a paper, you should identify how its argument is constructed. As discussed more in Chapter 4, consider constructing an evidence table that highlights the following elements as you read your sources:

- What was the overall argument of the study?
- Which experiments provide the strongest support for the argument? What were the methods and outcomes of these experiments? What assumptions were made in formulating the argument?
- What are the implications of the study?

It is not enough, however, to highlight the information: try to describe these elements in your own words. Doing so will help you differentiate your words from those used in the original source.

Keep Track of Your Sources

Given the amount of literature that you will read while developing research articles, literature reviews, and other biological writing, a solid note-taking system will help you remember who said what and where. Bibliographic software such as Mendeley and Zotero not only records your sources but also allows you to upload, search, and take notes on the .pdf versions of the papers.

The same software will help you later develop your Literature Cited section. Because many formats exist for listing references, it is important for you to fill in all the required fields for each source. The type of fields needed depends on the source (e.g., journal article, book chapter). Here are the most common fields required for referencing journal articles:

- Title
- Authors, first and last names
- Journal
- Year
- Volume
- Issue
- Pages, first and last

Thankfully, you can generally set up the software to automatically fill in the fields as you upload your papers. Information, however, can sometimes be extracted incorrectly (e.g., a missing page number, an incorrect volume), so you will want to double-check for accuracy.

Paraphrase and Compress Your Sources

In your English courses, when you were instructed on how to use sources in a paper, you were probably advised to use a mix of direct quoting, paraphrasing, and summarizing, with direct quotation the most common mode. Right now you may even be writing humanities and social science papers, where integrating direct quotes into your analysis will serve you well. But that is not the case in biology, where we almost never quote a source directly. Instead, we paraphrase.

Ken Hyland, who studied the citation practices of academics in different fields (Hyland 1999), attributes their differences in citation habits to the ways that knowledge is constructed in various disciplines. In the humanities, knowledge is created through direct engagement with previous ideas; the path to knowledge is iterative and recursive. In addition, the literature is more open to interpretation, so a writer's interpretation may not be shared by the audience. Directly quoting the literature in the humanities then helps the writer establish a common perspective with the reader. In contrast, research in the hard sciences follows a well-defined, more linear path: scientists cite sources in order to demonstrate how their findings fit into the network of accepted scientific facts. The facts are assumed to be correct. We might add that when scientists see a reference to a paper supporting something that they find dubious, they want to read the cited reference because frequently the supporting evidence cannot be summed up in a single quotation.

We described in Chapter 4 how you do not need to include all the details about an article in your literature review. Instead, you should compress your source, or cite only those details that are relevant for a particular point that you are making. To illustrate this point, let's focus on an abstract from Chapter 2 regarding the role of Ypel4 in the development of red blood cells. A complete summary might go something like this:

Lin et al. (2012) used a mouse model in order to determine the function of Ypel4 in erythropoiesis. They found that RNA expression significantly increases during the last stages of erythropoiesis, and that the protein is present in more highly differentiated blood cells. In addition, knocking down Ypel4 expression via RNA interference prevented expression of the Ter119 antigen and removal of the nucleus from the blood cells. Their findings suggest an essential role for Ypel4 in the development of red blood cells.

This summary nicely represents the purpose, methods, key results, and implication of the study, but such a summary is rarely needed in writing. A literature review on the Ypel family of proteins, for example, may cite the source in this way:

> Although many Ypel proteins are suggested to play a role in cell cycle regulation, Ypel4 was recently found to be necessary for the last stages of erythropoiesis (Lin et al., 2012).

The study could also be cited in a literature review on erythropoiesis:

> Genes involved in the last stages of red blood cell development include Ypel4 (Lin et al., 2012), . . .

Or even a review on Ter119:

> Expression of the Ter119 antigen is essential for terminal differentiation, and depends upon Ypel4 expression (Lin et al., 2012) . . .

For all of these contexts, a complete summary would be inappropriate. Instead, each of these sentences highlights a single detail. Such compression is common in biology writing. Note, however, the inclusion of the citation, which allows a reader to validate the information and learn more about the study on her own.

Paraphrasing Properly

Here is an excerpt from a published journal article (Buck and Axel 1991) followed by two attempts at paraphrasing. See if you can determine which paragraph is the appropriate paraphrase and which one crosses the line into plagiarism.

Original: The experimental design we employed to isolate genes encoding odorant receptors was based on three assumptions. First, the odorant receptors are likely to belong to the superfamily of receptor proteins that transduce intracellular signals by coupling to GTP-binding proteins. Second, the large number of structurally distinct odorous molecules suggests that the odorant receptors themselves should exhibit significant diversity and are therefore likely to be encoded by a multigene family. Third, expression of the odorant receptors should be restricted to the olfactory epithelium.

Attempt 1: Buck and Axel's cloning strategy to identify genes that encode for odorant receptors rested on three assumptions. First, the odorant receptors belong to a family of receptor proteins that transmit signals by interacting with G proteins. Second, the wide variety of odorants implies a great diversity among odorant receptors; therefore, odorant receptors are encoded by a family consisting of many genes. Finally, odorant receptors expression is limited to the olfactory epithelium.

Attempt 2: In designing a cloning strategy for odorant receptor genes, Buck and Axel (1991) assumed three things about the genes: first, they would be expressed only in the olfactory epithelium; second, they would be similar in sequence to those of other receptors that transmit signals via G proteins; and third, the genes would constitute a large, diverse family to reflect the wide variety of odorants.

Attempt 2 is the appropriate paraphrase. Attempt 1 would be considered plagiarism for two reasons. First, although the writer does well to mention Buck and Axel, he neglected to cite the actual article. Therefore, the reader will be unable to find the original source of the material discussed. Second, the excerpt is poorly paraphrased. The writer attempts to paraphrase by using synonyms to replace such terms as isolate, transduce, and GTP-binding proteins. The sentences, however, are structurally the same as the corresponding sentence in the original excerpt. Because a person's intellectual property consists of not only her ideas but also how those ideas are expressed, borrowing even the syntax of the original source material is considered plagiarism. You cannot simply cut a few words and swap in some synonyms and call that your paraphrase. A paraphrase must substantially change both the wording and structure of the source yet retain its essential meaning. Table 7.1 delineates the changes made in the second attempt that make a more successful paraphrase. Learning how to alter sentences will help you paraphrase better, and may even help clarify your own thoughts on the information.

TABLE 7.1 **Ways to Paraphrase Better**

Advice	Original Excerpt and First Paraphrase	Second Paraphrase
Change the subject.	Focus on odorant receptors	Focuses on genes
Convert active to passive verbs, or vice versa.	"odorant receptors . . . *are* restricted/encoded"	"genes . . . constitute"
Condense (e.g., convert clauses to phrases).	Second point: there is a lot of diversity among odorant receptors, *therefore* . . .	The genes constitute a large family *to reflect* the diversity
Change the order of ideas.	Order of assumptions is the same.	Order of assumptions is different.

Although we say that paraphrasing requires more than using a thesaurus, there are even occasions when a thesaurus does not help (e.g., a word or phrase is commonly used in the subfield or simply does not have a synonym). How does one go about determining this? Search for the term within literature reviews and research articles, and see if the author repeats the word or phrase or employs synonyms. You can also examine reviews and articles across a subfield to determine whether a term is unique to a particular scientist (and thus should be cited) or is commonly used in the subfield.

Remember that plagiarism is plagiarism whether you intended to do it or not. This means that learning to compress source material effectively through proper paraphrasing is among the most important moves that you need to master.

Do You Really Need to Cite that Source?

Students frequently wonder how to decide what information is so well known that no source needs to be cited. A good rule of thumb is to cite information that is known by less than 40% of your intended audience. This guideline doesn't mean that you have to poll potential readers, but you do need to characterize them. Writing for a journal with a broad audience, such as *Nature*, may require more citations than writing for a journal that is specific to a subfield, such as *The Journal of Neuroscience*. As you take more advanced courses or become more experienced in a research laboratory, you will become more familiar with what is known and not known in that particular subfield. In a college course, the instructor should specify the intended audience. We ask students in our class to write for other undergraduates who have taken the prerequisites for the course. So, for example, the audience would know what neuroreceptors are but probably would not be familiar with the neuroreceptor studied by the student-researcher.

Cite Sources of Illustrations and Images

As we noted earlier, you should not copy passages of text from an article and use them as direct quotations in your own paper. Yet biologists do sometimes copy figures they find in one source and use them in their own work, but when doing so they always properly credit the original source. They may even modify the image or its legend, especially if doing so makes the figure more suitable for the purpose at hand (while not misrepresenting or distorting the original).

In either case, the legend must identify both the source of the illustration and any modifications that you make. See Figure 4.4 for an example.

Cite Every Source with an In-Text Reference

All facts and ideas—whether they are expressed in images or text—must be credited to their sources. As we explained above, scientific writing tends not to use footnotes or endnotes for citing references. Instead, citations appear in the text. When preparing manuscripts, biologists use the format of the journal in which they plan to publish; such information can be found in the journal's Information for Authors. Check with your instructor regarding the preferred format for work in the course. If there is none, then choose the format of a specific journal, and stick to it.

In general, references are cited either by author name and publication year, or by a number. To illustrate these formats, we will use the recommendations of the Council for Science Editors (CSE) in Scientific Style and Format: The CSE Manual for Authors, Editors and Publisher, Eighth Edition (2014).

Name–Year System

The CSE name–year system of documenting sources consists of the last name of the first author (and the second if there are only two), as well as the year of publication. Citations can appear at the end of sentence or be incorporated in the sentence. Note how parentheses are used differently for these two formats:

Carson (1962) documented the environmental effects of the pesticide DDT.

DDT detrimentally affects not only the intended insects but also birds that ate the insects (Carson 1962).

← The use of parentheses depends upon whether the name–year citation appears within or at the end of a sentence.

This system has a couple of advantages because the citation is in the text, directly associated with the supported information. First, many readers can identify the paper as one that they have read, or would like to read, without having to turn to the Literature Cited section. In addition, the name–year system easily allows you to check that you are citing the correct source. The alternative system, discussed below, identifies references simply by numbers in the text; a typographical error, 5 for 15 for example, could send the reader searching for the incorrect paper.

Here are some things to keep in mind when citing using the name–year system:

Recognize All Authors
Remember that most papers have more than one author, so it is important to give credit to all the scientists. For papers with one or two authors, list the last name(s):

Woese and Fox (1977) compared the sequences of 16S ribosomal RNA from a large variety of organisms.

Sequence analysis revealed three domains of life (Woese and Fox 1977).

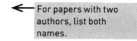 ← For papers with two authors, list both names.

For papers with more than two authors, list the last name only of the first author, and acknowledge the other writers with the phrase "et al." This phrase is short for "et alii," which means "and others" in Latin. Note that "et" does not take a period, but "al." does because the latter is abbreviated.

Hairston et al. (1960) proposed that the population of herbivores is limited more by predators than by the availability of food.

← For papers with more than two authors, use "et al."

The paper suggested that carnivores help maintain high levels of plant biomass (Hairston et al. 1960).

Use Punctuation to Separate References Within a Citation

Citations by different authors are ordered by publication year and separated by semicolons. In the example below, the three papers actually came from the same journal issue, so they are listed by page number.

The structure of DNA was first described in a trio of publications (Watson and Crick 1953; Wilkins et al. 1953; Franklin and Gosling 1953).

← Use semicolons to separate references, and list by publication year.

Distinguish Papers Published by the Same Author

Citations by the same first author are separated by commas and listed in chronological order:

Tetrahymena cell extracts were used to analyze how telomeres (chromosome ends) are synthesized (Greider and Blackburn 1985, 1987, 1989).

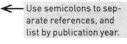 ← Use commas to separate publications by the same author.

Sometimes, a scientist may be super-prolific and publish multiple papers in a single year. In such cases, use lowercase letters to distinguish the papers:

A mathematical model for the action poten- ⟵ Publications from the
tial was described (Hodgkin and Huxley   same year are differen-
1952a, 1952b, 1952c, 1952d, 1952e).   tiated by letter.

Citation–Sequence and Citation–Name Systems

Sources can also be identified by assigned numbers in the text. Citation numbers appear as a superscript or within parentheses—before punctuation such as a comma or period. We have reproduced our Carson, 1962 example so that you can easily compare them.

DDT detrimentally affects not only the   ⟵ The 1962 book by
intended insects but also birds that ate the   Carson has been
insects[5].   replaced by a number.
                                                Note the space before
DDT detrimentally affects not only the   the superscript citation.
intended insects but also birds that ate
the insects (5).

The advantages of this documentation system are that it is simple and can be less distracting for the reader, especially if there are a lot of citations. There are two ways to assign numbers. In the citation–sequence system, sources are numbered in the order in which they first appear in the research article (if the source appears again later in the paper, the originally assigned number is reused). In the citation–name system, references are first alphabetized according to the last name of the first author and then assigned a number.

Numeric citations are separated by a comma, while consecutive citations can be written as the first and last numbers in the series connected by a hyphen.

Watson and Crick's structure of DNA was
informed by many studies.[15, 23–26]

> Use commas to separate citations, but hyphens when the citations are consecutive.

Citing Unpublished Information

Unpublished information is not listed in the Literature Cited section, yet still gets cited. Regardless of your citation system, sources of unpublished information appear in parentheses at the end of a sentence as shown below:

- (data not shown)
- (unpublished data)
- ([initials of author(s)], in press)
- ([initials of author(s)], in review)
- ([initials of author(s)], submitted)

Note that for a multi-authored paper, if any of these citations do not apply to all authors, add initials of those involved. For example, unpublished data from work of two of the three authors would be cited as follows: (MN and UL, unpublished data).

Alternatively, you can cite information as a "personal communication" after asking for permission. Be sure to specify the person in the citation by first initial and last name, and state the nature and source of the information. Note, too, that the citation is not in the Literature Cited but in the Acknowledgments section:

(2013 email from T. Jones to me; unreferenced, see "Acknowledgments")

> Include nature and source of information. Clearly state that the citation is not referenced.

Place Citations Where They are Most Relevant

When citing multiple sources, try to place your citations next to the most relevant information so that the reader does not have to determine the respective sources. The examples below use the name–year system, but the advice also applies to the citation–sequence and citation–name systems.

Needs Improvement

The use of double-stranded RNA to silence gene expression was first identified in nematodes but was soon applied to other model organisms as varied as fruit flies, zebrafish, and plants (Fire et al. 1998; Waterhouse et al. 1998; Kennerdell and Carthew 1998; Wargelius et al. 1999).

The citations are appropriately listed in chronological order, but the reader has to figure out how the sources correlate to the information.

Better

The technique of using double-stranded DNA to silence gene expression was first developed in nematodes (Fire et al. 1998) but was soon applied to other model organisms as varied as fruit flies (Kennerdell and Carthew 1998), zebrafish (Wargelius et al. 1999), and plants (Waterhouse et al. 1998).

Citations are closer to the respective information.

Finally, the location of the citation can help you minimize the number of times you cite. In the example below, the citation in the first sentence is implied to apply to the subsequent sentences.

Based on her studies of the chromosomes of *Zea maize* (corn), McClintock (1950) identified genetic elements that are able to jump within or between chromosomes. Some of these elements can move themselves, while others require activity of another element in

A citation at the beginning can be interpreted to apply to the information that follows.

order to move. These mobile elements are frequently associated with chromosome breakage and fusion, and can affect the expression of neighboring genes after moving to a new chromosomal site.

Prepare Your Literature Cited

The Literature Cited section (also called References or Cited References) lists all the cited references in an article—no more, no less. This section may be tedious to assemble, but it is a very important part of the research article. The Literature Cited section allows the reader to find and verify the information that you cite. Any typos or other inaccuracies in this section—including in such seemingly trivial information as a page number—will make readers question your authority and thoroughness as a researcher, especially if the citation leads to a completely unrelated paper. Therefore, be sure your Literature Cited section is both complete and accurate.

Tables 7.2 and 7.3 show the CSE templates for the most common types of sources included in a Literature Cited section. We refer you to *Scientific Style and Format: The CSE Manual for Authors, Editors and Publishers*, Eighth Edition (2014) for other sources not listed here.

The CSE reference format for citation–sequence and citation–name is the same, but the CSE reference format for the name–year system differs from both in placing the publication date immediately after the names of the authors. In addition, regardless of the type of publication (e.g., journal article, book, website), you should list the names of up to 10 authors; beyond that, use "et al." or "and others."

TABLE 7.2 **CSE Reference Format for Name–Year**

Type of Source	Template	Example
Journal article[1]	Author(s). Publication date. Article title. Journal title. Volume(issue):first and final pages.	Gould SJ, Lewontin RC. 1979. The spandrels of San Marco and the Panglossian paradigm: a critique of the adaptationist programme. Proc R Soc Lond B Biol Sci. 205(1161):581–598.
Book	Author(s). Publication year. Title. Edition. City of publication (state/country): publisher.	Darwin C. 1859. On the Origin of Species by Means of Natural Selection. London: Murray.
Chapter in an edited volume	Chapter author(s). Date. Chapter title. In: Book editor(s). Book title. Edition. City of publication (state/country): publisher. First and final pages of chapter.	Mayr E. 1954. Change of genetic environment and evolution. In: Huxley J, Hardy AC, Ford EB, editors. Evolution as a process. London: Allen and Unwin. p. 157–180.
Lab manual or course handout	Author(s). Publication year. Title. City of publication (state/country): publisher.	Schwartz TU. 2008. 7.02 laboratory manual. Cambridge (MA): Massachusetts Institute of Technology.
Website[2]	Homepage title. Date of publication. Edition. City of publication (state/country): publisher; <date updated; date accessed>. Notes including URL.	Information about chronic fatigue syndrome (CFS). Atlanta (GA): Centers for Disease Control and Prevention (US); [updated 2012 May 14; accessed 2014 Jan 15]. http://www.cdc .gov/cfs/general/index.html

[1]For publication date, include the season or month and day when the journal does not have a volume or issue number. Journal names are abbreviated according to the guidelines of the International Organization for Standardization (ISO).
[2]Choose your website wisely. Because most information on the internet constantly changes and is not peer-reviewed, try to rely only websites that are recognized authorities (e.g., end with ".edu," ".gov," or ".org"). Publication dates can sometimes be hard to locate on websites, but you should at least include the dates (including month and day) when the information was updated and accessed.

TABLE 7.3   CSE Reference Formats for Citation–Sequence and Citation–Name

Type of Source	Template	Example
Journal article[1]	Author(s). Article title. Journal title. Publication date; volume (issue): first and final pages.	Gould SJ, Lewontin RC. The spandrels of San Marco and the Panglossian paradigm: a critique of the adaptationist programme. Proc R Soc Lond B Biol Sci. 1979;205(1161):581–598.
Book	Author(s). Title. Edition. City of publication (state/country): publisher; publication year.	Darwin C. On the Origin of Species by Means of Natural Selection. London: Murray; 1859.
Chapter in an edited volume	Chapter author(s). Chapter title. In: Book editor(s). Book title. Edition. City of publication (state/country): publisher; publication year. First and final pages of chapter.	Mayr E. Change of genetic environment and evolution. In: Huxley J, Hardy AC, Ford EB, editors. Evolution as a process. London: Allen and Unwin; 1954. p. 157–180.
Lab manual or course handout	Author(s). Title. City of publication (state/country): publisher; publication year.	Schwartz TU. 7.02 laboratory manual. Cambridge (MA): Massachusetts Institute of Technology; 2008.
Website[2]	Homepage title. Date of publication. Edition. Place of publication: publisher; <date updated; date accessed>. Notes including URL.	Information about chronic fatigue syndrome (CFS). Atlanta (GA): Centers for Disease Control and Prevention (US); [updated 2012 May 14; accessed 2014 Jan 15]. http://www.cdc.gov/cfs/general/index.html

[1] For publication date, include the season or month and day when the journal does not have a volume or issue number. Journal names are abbreviated according to the guidelines of the International Organization for Standardization (ISO).
[2] Choose your website wisely. Because most information on the internet constantly changes and is not peer-reviewed, try to rely only websites that are recognized authorities (e.g., end with ".edu," ".gov," or ".org"). Publication dates can sometimes be hard to locate on websites, but you should at least include the dates (including month and day) when the information was updated and accessed.

References are Alphabetized by the Last Name of the First Author

The only exception is if you use the citation–sequence system for documentation. There are additional things to keep in mind as you list your sources. The examples below use the CSE reference format for the name–year system.

Distinguish References Written by First Authors with the Same Last Name

Articles whose first authors share the same last name are listed alphabetically by first initial.

> King JL, Jukes TH. 1969. Non-Darwinian evolution. Science. 164(3881):788–798.

> King MC, Wilson AC. 1975. Evolution at two levels in humans and chimpanzees. Science. 188(4184):107–116.

Papers with the same first author are listed in chronological order.

> Ehrlich PR, Raven PH. 1964. Butterflies and plants: a study in coevolution. Evolution. 18(4):586–608.

> Ehrlich PR, Raven PH. 1969. Differentiation of populations. Science. 165(3899):1228–1232.

If more than one paper was published in the same year by the same first author, use letters to distinguish between them, and list them according to exact publication date or page number (if the articles are in the same issue of a journal).

> Hodgkin AL and Huxley AF. 1952a. A quantitative description of membrane current and its

application to conduction and excitation in nerve. J Physiol. 117(4):500–544.

Hodgkin AL and Huxley AF. 1952b. Propagation of electrical signals along giant nerve fibers. Proc R Soc Lond B Biol Sci. 140(899):177–183.

Align Your References

Format each entry in your reference list either with the whole text flush against the left margin, or with a hanging indent. For example, the Ehrlich and Raven references above were flush, while the Hodgkin and Huxley references were listed with a hanging indent.

Reference Format Depends Upon the Journal

While the CSE is the most common documentation system and will serve most students well, you should know that if you aim to publish your work, you will discover there is no single format that is used by all journals. Compare the formats for a landmark biology paper when it appears in Literature Cited sections of four widely read journals:

Cell
Avery, O.T., Macleod, C.M., and McCarty, M. (1944). Studies on the chemical nature of the substance inducing transformation of Pneumococcal types: Induction of transformation by a desoxyribonucleic acid fraction isolated from Pneumococcus Type III. J. Exp. Med. 79, 137–158.

Nature

Avery, O. T., Macleod, C. M. & McCarty, M. Studies on the chemical nature of the substance inducing transformation of Pneumococcal types: Induction of transformation by a desoxyribonucleic acid fraction isolated from Pneumococcus Type III. *J. Exp. Med.* **79,** 137–158 (1944)

Science

O. T. Avery, C. M. Macleod, M. McCarty, Studies on the chemical nature of the substance inducing transformation of Pneumococcal types: Induction of transformation by a desoxyribonucleic acid fraction isolated from Pneumococcus Type III. *J. Exp. Med.* **79,** 137–158 (1944).

Proceedings of the National Academy of Science (U.S.A.)

Avery OT, Macleod CM, McCarty M (1944) Studies on the chemical nature of the substance inducing transformation of Pneumococcal types: Induction of transformation by a desoxyribonucleic acid fraction isolated from Pneumococcus Type III. *J Exp Med* 79(2):137–158.

It can be maddening to keep track of minute differences in the formatting of author names and journal volume, or in the number of authors listed before "et al." is added. However, if you use bibliographic software (e.g., Mendeley, Zotero, discussed above), you can keep track of your sources and can fairly quickly convert references from one template to another.

SOURCE CHECKLIST

Do

✓ Keep track of every source.
✓ Paraphrase accurately but not too similarly to the original text.
✓ Select a format for citations and references and use it consistently.
✓ Be complete and accurate in your Literature Cited section.

Don't

✓ Take credit for work that is not your own.
✓ Use direct quotes: paraphrase instead.
✓ Cite a source that does not appear in the Literature Cited section, or list a source in the Literature Cited that is not cited in the manuscript.

REVIEWING LIKE A BIOLOGIST

Peer review is the gold standard used by biological journals when they are deciding to accept or reject scientific research papers for publication. A research paper that is submitted to a journal is evaluated by an editor, who then sends it out to two or more experts in the field. These referees prepare anonymous reviews on the strengths and weaknesses of the manuscript, advising the editor whether to accept it for publication and giving authors suggestions for improving the manuscript.

The two reviews are sent to the editor. If the two referees are essentially in agreement, the editor will decide whether to accept or reject the manuscript. If the referees disagree on important points, the editor will request one or more additional reviews. When the final decision on the manuscript is settled, the editor writes the author to explain the decision and forwards copies of the anonymous reviews. End of story? No! Now the scene shifts to the author's laboratory.

Receiving a Review

Let's say that you are the author of the paper that we are following. Before you open the email from the editor, think about a couple of things. First, students are usually surprised when their carefully polished manuscript comes back with editorial criticism; however, as you become more experienced, you realize that you, too, find things to improve when you look at a manuscript several days after you "finished" it. Second, these reviews give you the advice of two experts in the field that will help you improve the manuscript, whether or not it is accepted by this journal. No doubt you had people at your institution comment on your drafts, but those readers have probably heard you talk about the project from time to time. The journal's referees will be from other institutions. Viewing the work for the first time, they will be more likely to spot places where the story is hard to follow than neighbors familiar with your work.

The manuscript may be accepted as written (a rare event); it may be accepted if you can respond to the referees' concerns; or it may be rejected. In any case the editor's letter and the reviews should give clear explanations of the concerns. These are the basis for your next move. If the paper is accepted pending some revisions, the choice is easy. Usually many of these concerns involve improving text and/or illustrations, and you'll probably find that some, if not all, do improve the paper. However, you may think the referees are mistaken in some concerns and decide either to make no change or to add text to explain your original point more effectively. When you return the revised manuscript to the journal, you add a letter to the editor, thanking the referees and explaining point by point how you have modified the manuscript and giving your reasons for not accepting some of the referees' suggestions.

If the paper has been rejected, your decision is more complicated. You can either reformat the paper for a different journal or try sending it with a rebuttal letter to the original journal. In either case, responding to the concerns of the original reviews should help you to have a better manuscript.

Writing a Review

Even though it may be a while before you find yourself in the position of needing to referee a paper for a journal, it is useful to learn to write a good review. Why? Reviewing the final draft of your own paper is a good way to stand back and see what a reader will think of it. It is another way of improving your writing. Your professor or TA may also require you to do peer review ("peer" in this case means your classmates), and while this may seem like a simple process of exchanging papers and commenting on them, in fact there are better and worse ways of performing peer review, whether in a classroom or in a professional setting. Here is our advice: allow your final draft to sit for a few days before you review it. Then ignore your title and abstract. Begin with the Introduction and read the paper, trying to see it with fresh eyes.

Begin your review with a paragraph saying concisely what you think the paper's major findings are, how well they are supported, and what their significance is. This initial paragraph helps the editor decide on the merit of the paper and also whether it is appropriate for the readers of a particular journal (e.g., if it is an analysis of the structure of a particular protein, is there a reason that this particular analysis will be of interest to readers of a journal of genetics?). This paragraph also helps the author see how well the reviewer has understood the manuscript and where the manuscript could be improved (see the sample peer review later in the appendix).

After this overall evaluation, there are specific points to consider. If your evaluation is negative, you must support this with clearly defined concerns. If you would support publication after some revision—perhaps another experiment or discussion of another possible interpretation of the data—the desired changes should be described. Even for papers that you strongly support, if you see anything that could be improved or have a suggestion that the authors may not have thought of, these can be included. For this last category, it is important to make clear what suggestions are strictly optional.

After the opening paragraph, all specific points are listed separately and numbered so that, in the letter that accompanies a revised manuscript, the author can respond point by point. Questions you should consider include the following:

- Do the title and abstract accurately represent the contents of the manuscript?
- Does the Introduction provide the relevant background information and lead directly to the goal of the research? Are the contributions of others fairly acknowledged?
- Are the experimental techniques appropriate for the project? Are they adequately described?
- Are the critical results appropriately displayed? Do the figures and tables support the conclusions? Are there other possible interpretations of the data that should be tested experimentally? Answering this last question requires some restraint: suggest only experiments that are crucial to the conclusion of the manuscript being reviewed. Even if you are suggesting "only one more" experiment, consider carefully whether the information gained justifies the delay in publishing the paper.
- Does the Discussion section link the experimental results to the goal of the research provided in the Introduction? Does the Discussion relate this study to other work in the field in an insightful way? Is any speculation about the impact of the results justified?
- Are the references relevant and complete?
- Can the text be shortened? Are there any unnecessary figures or tables?

When you are asked to review papers for others, there are several additional things to keep in mind:

- You are expected to consider the manuscript a privileged document, not to be discussed with others—including people in your lab.

- You should agree to review a paper only if you have no conflict of interest and can evaluate it with a positive, impartial attitude.
- You are writing a review of the manuscript, not the work, and your goal is to ensure effective communication of solid science. The authors will receive a copy of your review, which should be written with this in mind. If you have special comments for the editor, these can be sent separately.
- You are not expected to correct grammar, spelling, and so forth.

Sample Peer Review

The nuclear RNA-binding protein, Hrp84, has been found to offer some protection against damage to neurons in culture. This manuscript describes pioneering experiments to use a genetically engineered mouse (LGD3) to test whether this protection could improve motor function in a mouse carrying a mutation responsible for many human cases of ALS (Lou Gehrig's disease). LGD3-M and LGD3-N mice also carried extra copies of the gene for Hrp84 that could be specifically activated in either muscle (LGD3-M) or nerve (LGD3-N) cells. Although severely limited by the time available, the study provides suggestive evidence that hrp84 expression in nerves, but not muscle, may be able to provide some protection from symptoms of ALS and thus be a target for treating ALS in the future.

I have only minor comments about clarifying some of the methods and figures:

1. Consider adding more details on how the clinical score for the different mice is assayed. In the Methods, clinical score is mentioned as being measured by visual inspection. I wondered how you can measure scores like 1.17 based on your eye? Or are these numbers the result of averaging? Basically, I wondered how many significant figures you can measure based on eye alone (how can you differentiate 1.17 from 1.18, for example). Since a lot of your results depend on clinical score, it might be useful to clarify this.
2. Sometimes you used phrases like "not all data points are significant" or "LGD3-N mice overexpressing Hrp84 in nerves showed significantly fewer disease symptoms than LGD3-M mice," but

I wasn't able to find any statistical tests of significance in your text or figures. You might want to be careful about using phrases like "significant" so people don't confuse it with statistical significance. Or clarify in what sense you mean the data are "significant."

3. I wondered why you didn't show grip strength of wild-type and transgenic mice not overexpressing Hrp84 in Figure 1C and D. If you have the data, it might be a good idea to show these controls to make panels C and D consistent with the rest of the figure. Or mention why wild-type and transgenic mutant animals were not included, as you did in Figure 2.

4. In Figure 4C, the caption says average grip strength is improved in LGD3-N mice, but from the bar plot it looks like average grip strength is lower in LGD3-N mice than in LGD3-M mice. Since the trend in the rest of the figure is the opposite, you might want to explain in more detail how grip strength is improved in LGD3-N mice.

CHECKLIST FOR WRITING A REVIEW

Do

✓ Review manuscripts with a positive, impartial attitude and with the goal of improving both the science and the writing.

✓ Start with a summary of the manuscript's key data and impact.

✓ Be specific when explaining your concerns and suggestions.

Don't

✓ Suggest additional experiments for their own sake: the additional information must justify the delay of publication.

✓ Focus on issues of grammar or spelling unless this was explicitly requested (most assume that this will happen at the *copyediting* stage, not the peer review stage).

✓ Discuss the manuscript with other people.

REFERENCES

Chapter 1

Bean, John. 2011. *Engaging Ideas: The Professor's Guide to Integrating Writing, Critical Thinking, and Active Learning in the Classroom.* San Francisco: Jossey-Bass.

Chapter 2

Avery, Oswald T., MacLeod, Colin M., and Maclyn McCarty. 1944. "Studies on the Chemical Nature of the Substance Inducing Transformation of Pneumococcal Types: Induction of Transformation by a Desoxyribonucleic Acid Fraction Isolated from Pneumococcus Type III." *Journal of Experimental Medicine* 79(2): 137–58.

Halloran, S. Michael. 1984. "The Birth of Molecular Biology: an Essay in the Rhetorical Criticism of Scientific Discourse." *Rhetoric Review* 3(1): 70–83.

Watson, James D., and Francis Crick. 1953a. "Molecular Structure of Nucleic Acids: A Structure for Deoxyribose Nucleic Acid." *Nature* 171(4356): 737–8.

Watson, James D., and Francis Crick. 1953b. "Genetical Implications of the Structure of Deoxyribonucleic Acid." *Nature* 171(4361): 964–7.

Watson, James D., and Francis Crick. 1953c. "The Structure of DNA." *Cold Spring Harbor Symposia on Quantitative Biology* 18: 123–31.

Chapter 3

Dulbecco, Renato. 1949. "Reactivation of Ultra-violet-inactivated Bacteriophage by Visible Light." *Nature* 163(4155): 949–50.

Chapter 4

Meselson, Matthew, and Franklin W. Stahl. 1958. "The Replication of DNA in *Escherichia coli*." *Proceedings of the National Academy of Sciences of the United States of America* 44(7): 671–82.

Chapter 5

Alley, Michael. 2013. *The Craft of Scientific Presentations: Critical Steps to Succeed and Critical Errors to Avoid*. New York: Springer.

Reynolds, Garr. 2010. *The Naked Presenter: Delivering Powerful Presentations With or Without Slides*. San Francisco: New Riders.

Reynolds, Garr. 2011. *Presentation Zen: Simple Ideas on Presentation Design and Delivery*. San Francisco: New Riders.

Tufte, Edward R. 1983. *The Visual Display of Quantitative Information*. Cheshire: Graphic Press.

Williams, Robin. 2014. *Non-Designer's Design Book*. San Francisco: Peachpit Press.

Chapter 6

Braddock, Richard. 1974. "The Frequency and Placement of Topic Sentences in Expository Prose." *Research in the Teaching of English* 8(3): 287–302.

Gopen, George, and Judith Swan. 1990. "The Science of Scientific Writing." *American Scientist* 78(6): 550–8.

Hood, Thomas. 1837. "Copyright and Copywrong." *The Athenaeum*, April 22: 285–7.

Chapter 7

Avery, Oswald T., MacLeod, Colin M., and Maclyn McCarty. 1944. "Studies on the Chemical Nature of the Substance Inducing Transformation of Pneumococcal Types: Induction of Transformation by a Desoxyribonucleic Acid Fraction Isolated from Pneumococcus Type III." *Journal of Experimental Medicine* 79(2): 137–58.

Buck, Linda, and Richard Axel. 1991. "A Novel Multigene Family May Encode Odorant Receptors: A Molecular Basis for Odor Recognition." *Cell* 65(1): 175–87.

Chargaff, Erwin, Lipshitz, Rakoma, Green, Charlotte, and Marion E. Hodes. 1951. "The Composition of the Deoxyribonucleic Acid of Salmon Sperm." *Journal of Biological Chemistry* 192(1): 223–30.

Chargaff, Erwin, Lipshitz, Rakoma, and Charlotte Green. 1952. "Composition of the Desoxypentose Nucleic Acids of Four Genera of Sea-urchin." *Journal of Biological Chemistry* 195(1): 155–60.

Hyland, Ken. 1999. "Academic Attribution: Citation and the Construction of Disciplinary Knowledge." *Applied Linguistics* 20(3): 341–67.

CREDITS

Page 11, *Figure 2.1* (left): © A Barrington Brown, Gonville and Caius College/Science Photo Library.

Page 11, *Figure 2.1* (right): Courtesy of the Rockefeller Archive.

Page 13, *Figure 2.2*: © "Time" from Flickr, © Alexander Boden, https://www.flickr.com/photos/bogenfreund/. Reprinted under Creative Commons 2.0 license.

Page 23, *Figure 2.7*: © Eun Young Choi.

Page 24, *Figure 2.8*: © Jihye Kim.

Page 25, *Figure 2.9*: Reprinted by permission of Anne Huang.

Page 26, *Figure 2.10*: Reprinted by permission of Kamena Kostova.

Page 101, *Figure 3.2*: Reprinted by permission of Ashley Funk.

Page 148, *Figure 5.1*: "Eureka!" from Flickr, © ARG_Flickr, https://www.flickr.com/photos/arg_flickr/8795938525/. Reprinted under Creative Commons 2.0 license.

Page 156, *Figure 5.4*: © Jan Van Aarsen.

Page 158, *Figure 5.5*: © Jan Van Aarsen.

Page 160, *Figure 5.6*: © Christopher Bird.

Page 164, *Figure 5.8*: Reprinted by permission of Xin Xin.

INDEX